D0138290

MAIN

REFERENCE

# ATLAS OF THE WORLD'S DESERTS

# ATLAS OF THE WORLD'S DESERTS

Nathaniel Harris

**Fitzroy Dearborn**
An Imprint of the Taylor and Francis Group
New York • London

Published by
Fitzroy Dearborn
An imprint of the Taylor and Francis Group
29 West 35th Street
New York, NY 10001-2299

*and*

Fitzroy Dearborn
An imprint of the Taylor and Francis Group
11 New Fetter Lane
London EC4P 4EE

British Library and Library of Congress Cataloging-in-Publication Data are available

ISBN 1-57958-310-5

For The Brown Reference Group plc

Editors: Robert Anderson, Shona Grimbley, Sally McFall, Ben Morgan, Henry Russell
Designer: Lynne Ross
Cartographer: Darren Awuah
Picture Research: Becky Cox
Additional Text: Steve Parker
Production Manager: Matt Weyland
Production Director: Alastair Gourlay
Managing Editor: Tim Cooke
Indexer: Kay Ollerenshaw
Editorial Director: Lindsey Lowe

This edition first published by
Fitzroy Dearborn, an Imprint of the Taylor and Francis Group 2003

Printed in Hong Kong.

PICTURE CREDITS

Art Archive: British Library 139, Musée d'Orsay/Dagli Orti 147; Bruce Coleman Collection: Jen & Des Bartlett 104, E. Bjurstrom 88, Fred Bruemmer 103, John Cancalosi 90, 99, Bruce Coleman Inc. 87, 120t, 120b, Jules Cowan 117, 76, 82, M.P.L. Fogden 80b, 84t, 101, Jeff Foott 47b, 78, 81, Tore Hagman 79, 83, HPH Photography 108, P. Kaya 110, Dr Eckart Pott 80t, 84b, Kim Taylor 97, 98; Corbis: 20, Tom Bean 155, Richard Cummins 74, Robert Garvey 159, Raymond Gehman 113, Richard Hamilton-Smith 50, Peter Johnson 48, Wolfgang Kaehler 144, Steve Kaufman 14, David Lees 133, Charles Lenars 17, Peter Lillie 57, Neil Rabinowitz 47t, Galen Rowell 19, Paul A. Souders 8, Space Shuttle Endeavor 52, Gordon Whitten 12, 46, Martin Withers 49; Hutchison Library: 53, Dave Brinicombe 25, 172, O.R. Constable 170, H.R. Dorig 123, Nancy Durrell Mckenna 150, Robert Francis 126t, Mary Jelliffe 137, 163, Michael Kahn 36, Brian Moser 72, 73, Stephen Pern 59, Bernard Regent 173, Andre Singer 157, Andrew Sole 177, Isabella Tree 33, 65, Audrey Zvoznikov 62, 63; Image Bank: Harald Sund 142, Jose Szkodzinski 143; Library of Congress: 184; NHPA: A.N.T. 96, A.N.T./Ern Mainka 94, Anthony Bannister 105, 91, Robert Erwin 111, Pavel German 100, Daniel Heuclin 95b, 106, Hellio & Van Ingen 95t, Lady Philippa Scott 85; Robert Hunt Library: 135, Black Star 141; Science Photo Library: Tony Buxton 55, Bernard Edmaier 42, NASA 18, Sinclair Stammers 15; South American Pictures: Chris Sharp 127; Still Pictures: Adrian Arbib 37, 154, Romano Cagnon 182, Chris Caldicott 132, 153, William Campbell 152, Mark Edwards 178, Xavier Eichaker 92, Michel Gunter 28, 44, John Isaac 171, Emmanuel JeanJean 23, Klein/Hubert 54, 149, Gerard & Margi Moss 129, Gil Moti 183, Stephen Penn 21, Kevin Schafer 175, Jorgen Schytte 185, Roland Seitre 22, 109, 126b, 162b, 162t, VOLTCHEV-UNEP 30, Gunter Ziesler 125; Sylvia Cordaiy Photo Library: Dorothy Burrows 116, David William Gibbons 121, Gable 181, Johnathan Smith 168; Travel Ink: Allan Hartley 130.

# CONTENTS

| | | |
|---|---|---|
| | **Introduction** | 6 |
| | *Atlas: World Map of Aridity* | 10 |
| CHAPTER 1 | How Deserts Form | 12 |
| | *Atlas: African Deserts* | 24 |
| CHAPTER 2 | Sand, Rock, and Rubble | 42 |
| | *Atlas: Asian Deserts* | 58 |
| CHAPTER 3 | Plants of the Desert | 74 |
| CHAPTER 4 | Creatures of the Desert | 88 |
| | *Atlas: American Deserts* | 112 |
| CHAPTER 5 | The Desert in History | 130 |
| CHAPTER 6 | The Modern Desert | 144 |
| | *Atlas: Australia and the Poles* | 156 |
| CHAPTER 7 | Wealth from the Desert | 168 |
| CHAPTER 8 | Spreading Deserts | 178 |
| | *Glossary* | 186 |
| | *Bibliography* | 188 |
| | *Index* | 190 |

***Left*** **Limestone columns rise from the Pinnacle Desert in Western Australia. The hardened columns, which have been exposed by weathering in this coastal region, range from only a few centimeters to 5 meters (16 ft.) in height.**

# INTRODUCTION

In the Western imagination the word "desert" most often evokes a landscape of endless gigantic sand dunes, dazzling white under a cloudless hot-blue sky and a blazing sun. This landscape of the imagination is likely to be empty – deserted – except, perhaps, for a caravan of nomads and camels that inches slowly across the horizon, or a lone man stumbling, sun-blackened and sun-parched, through the heat haze. Or there may even be an emerald-green oasis, where tents are set out in the shade of a palm grove – though this, of course, may be nothing but a tantalizing mirage. This is the magnificent and exotic landscape of movies such as David Lean's *Lawrence of Arabia* (1962) and Bernardo Bertolucci's *The Sheltering Sky* (1990), and of countless adventure stories of intrepid travelers and explorers.

This idealized or classic landscape is not pure fantasy: parts of the Sahara, Arabian, and other deserts fit quite well with this image – though perhaps with less Technicolor vibrancy. The stereotype does, however, contain some misleading notions, of which the most notable is that all deserts are hot, and that heat is crucial in defining what constitutes a desert. Temperature actually plays a secondary role or no role in such definitions – not all deserts are hot, and even so-called hot deserts are not hot all the time. The Gobi Desert deep within Central and East Asia, for example, has relatively cool but erratic temperatures even in summer and can be brutally cold in winter, and in the Sahara temperatures can easily plummet to 4°C (39°F) at night. Modern geographers also recognize the category of the polar desert, applying it to all of Antarctica and parts of the Arctic (notably Greenland), where temperatures day and night stand at the opposite extreme to those of daytime hot deserts.

Even a brief perusal of the photographs included in this book will suggest a much more varied, and even nebulous, notion of what is – or is sometimes – meant by the term "desert." There are vast gravel plains, gleaming expanses of sun-baked salt, and rugged, eroded landscapes of pinnacles, canyons, and rock arches. There are deserts smothered with flowers and blooming cacti; there are others studded with oil wells or scarred by quarries. Some are washed by the ocean and bathed in fog, and some are ice-encrusted polar wildernesses. One of the surprising facts encountered in this book is that only 20 to 30 percent of the world's deserts are covered by sand, and that the world's great deserts in fact encompass a huge variety of terrains, not only relative to each other but sometimes within their own boundaries. There is, moreover, little exotic about the desert biome – almost 20 percent of the earth's land surface is desert, and there are deserts in almost every continent and at every latitude. Of the continents only Europe has no desert area. For many peoples of the world the desert is not a remote fantasy but a reality that impinges on their everyday lives.

## Defining the desert

Definitions of the term "desert" are neither static nor absolute. All over the world the term "desert" and its foreign-language equivalents are culturally and topographically specific. European words such as "desert," "*désert*," and "*Wüste*" emphasize the sense of abandonment that is the standard Western response to the desert landscape – an idea that is also reflected in the etymology of the name of the Namib Desert in southern Africa – "the place where there is nothing." Arabic has not one but several words for "desert," including *erg* (applied to large areas of sand or "sand seas") and *hammada* (applied to stony plains), as well as the more general *sahra*, from whose plural form – *sahara* – the world's largest desert takes its name. The Turkic *kum* means literally "sand," reflecting the sandy wastes of Central Asia – hence the Kara-Kum, or "Black Sand," of Turkmenistan and the Kyzyl-Kum, or "Red Sand," of neighboring Uzbekistan and Kazakhstan – while the Persian *dasht* means "plain" as well as "desert," in reference to the plateau deserts that dominate central Iran.

Physical geographers and geologists must at least attempt to be more scientific in their definitions of what constitutes a desert, and they have debated and extended the possible meanings. Today they agree that the key determining factor is aridity, or the lack of plentiful and consistent rainfall – generally defined as less than 250 millimeters (10 in.) of annual precipitation. Such a definition extends the meaning of desert well beyond its traditional confinement to the hot deserts that have so exercised the European imagination. As Chapter 1 shows, low rainfall is a characteristic not only of the subtropical regions where most of the hot deserts – the Sahara, Arabian, and Australian deserts, for example – are located, but also of continental interiors, the western sides of continents, the leeward side of high mountain ranges, and parts of the Arctic and Antarctic regions.

Even this definition is by no means watertight; strict definitions always create seeming anomalies. The Kalahari in southern Africa is labeled a desert in every atlas, and its very name – meaning "the Great Thirst" – would appear to confirm this status. But most of the Kalahari receives roughly twice the amount of the annual maximum allowable precipitation and has a relatively rich vegetation, and therefore for some scholars this would-be desert falls outside the strict definition of the term. However, more complex definitions of aridity take into account the rate of evaporation as well as the amount of precipitation, and the Kalahari, despite its rainfall, has little standing water due to the dry heat that rapidly evaporates much of the land's moisture. In their pursuit of exact definitions, experts have sometimes devised formulas to indicate a particular region's "Index of Aridity." One of the simplest, the Lang Rain Factor, for example, divides the annual precipitation (in millimeters) by the mean annual temperature (in centigrade).

Other arid regions, while not generally called deserts and often receiving slightly more than the regulation 250 millimeters (10 in.) of rainfall, display some of the characteristics of deserts. Such borderline "semiarid" regions are often covered by the terms "semidesert" or "drylands." The Sahel in sub-Saharan Africa is one important area of semidesert. In recent years this vast region has come under close scrutiny as its poor but locally crucial arable and pastoral lands have become degraded and the Sahara Desert has crept southward.

## The living landscape

Surprisingly perhaps, water plays a key role in shaping the desert terrain. This is because, when water does finally make its appearance in the desert, it usually does so in torrential form – powerful, destructive floods that rip through the land, sweeping away any debris or loose vegetation and over the centuries cutting channels – called "wadis" in North Africa and Arabia and "arroyos" in the Americas – deep into the landscape. Despite appearances, deserts are often mobile, changing landscapes, uniquely vulnerable to the often dramatic metamorphoses worked by weathering agents such as water, heat, and wind. Sand dunes slowly shift and grow; glistening salt pans become lakes and then dry hard again within weeks or days; and over millennia rocks are scoured and eroded into dramatic or bizarre forms, such as flat-topped mesas, mushroom-shaped zeugens, and awe-inspiring rock arches. The metamorphoses of the desert terrain form the subject of Chapter 2.

## Life in the desert

Conventional wisdom depicts the desert as almost devoid of vegetation or wildlife, save perhaps for a sidewinding snake or rearing scorpion. It is seen as abandoned by human beings, who in this hostile environment are thought of as interlopers or aliens, there only because they are on their way to somewhere else or because they have fatally lost their way. In Chapters 3, 4, and 5 we shall see how many deserts, despite their dearth of water – the precondition for the survival of life – in fact provide a remarkably

*Overleaf* **Map showing the world's aridity zones – yellow areas indicate regions with hyperarid or arid levels of precipitation. The world's major deserts and the atlas pages devoted to them in this book are also shown.**

fertile habitat for plants, animals, and humans alike, each of which have found ingenious ways of making the best of the desert. Plants store water through months of drought or blossom and seed after rare rainfall in a matter of days, transforming bare landscape into dazzling fields of color. Animals live by night or burrow deep underground, or – as in the case of reptiles – are physiologically adapted to withstand the desert's temperature extremes. Humans living in and on the fringes of deserts have developed unique lifestyles that usually feature nomadism – a fluid way of life that is able to adapt swiftly and creatively to the vicissitudes of this harsh environment. Some of the world's earliest civilizations – including those of ancient Egypt and Mesopotamia – formed on the margins of great deserts, where the strenuousness of life called for the utmost in human endeavor.

## The changing desert

Traditional nomadism is in most deserts a dying way of life. Colonialism, urban-biased political structures, and modern lifestyles and technologies derived from the West have proved antipathetic to traditional peoples everywhere, and the peoples of the desert are no exception. In Chapters 6, 7, and 8 we shall see how the desert, like every other biome, is facing unprecedented challenges to its very survival. In the late 19th and 20th centuries many deserts were found to harbor important mineral reserves, most significantly oil and natural gas, and consequently were and continue to be subject to intensive exploitation.

Perhaps more damaging still has been the ambition to "make the deserts bloom" – to irrigate formerly arid land, often by exploiting nearby rivers or the water table. The Negev Desert in Israel, the Kara-Kum in Turkmenistan, and the Libyan Desert have all been subject to grandiose and often ill-considered schemes motivated by a mixture of political or propagandistic concerns as well as by more humanitarian considerations. The effects of such programs have often been catastrophic, best exemplified by the demise of the Aral Sea following the development of the Kara-Kum Canal during the Soviet period. Of more general environmental concern is the global problem of desertification – the degradation of the semidesert lands and drylands that border established, "natural" deserts through poor agricultural practices – a development that threatens millions of people with poverty and hardship. It is with this looming global catastrophe that the final chapter of this book is concerned.

## The desert atlas

Interspersed with the text chapters of the book are atlas sections, arranged broadly by continent. All the world's great deserts are represented, including the polar deserts of the Arctic and Antarctic, and are richly annotated in order to give the reader a detailed understanding of their geography, ecology, and history. The maps are oriented to the north and, in addition to physical features – such as mountains, rivers, areas of sand, cities, towns, major roads, and railroads – include major political features such as national borders. With the exception of the United States and Australia, however, provincial or state borders are not shown. At the end of the book a glossary defines semispecialist terms used in the text. There is also a bibliography of the sources used for this book and recommended further reading for the reader who wishes to pursue a particular subject more deeply. The references include recent scientific and scholarly publications and websites; the latter are particularly useful for up-to-date information on changing data, such as rates of desertification.

This book draws on current scholarship, but it is not aimed at specialists. Instead it aims to give the general reader an insight into one of the world's least known habitats. The environmental concerns of recent years have tended to cluster around more appealing habitats, leaving the desert – too often perceived as some kind of ecological vacuum – largely overlooked. In its small way this book may help to rectify this injustice.

The Arctic
pp. 164–165
Great Basin
pp. 114–117
Mojave/
Sonoran
pp. 118–121
Sahara
pp. 26–31
Sechura
pp. 122–123
Atacama
pp. 124–127
Namib
pp. 32–33
Patagonia
pp. 128–129
The Antarctic
pp. 166–167

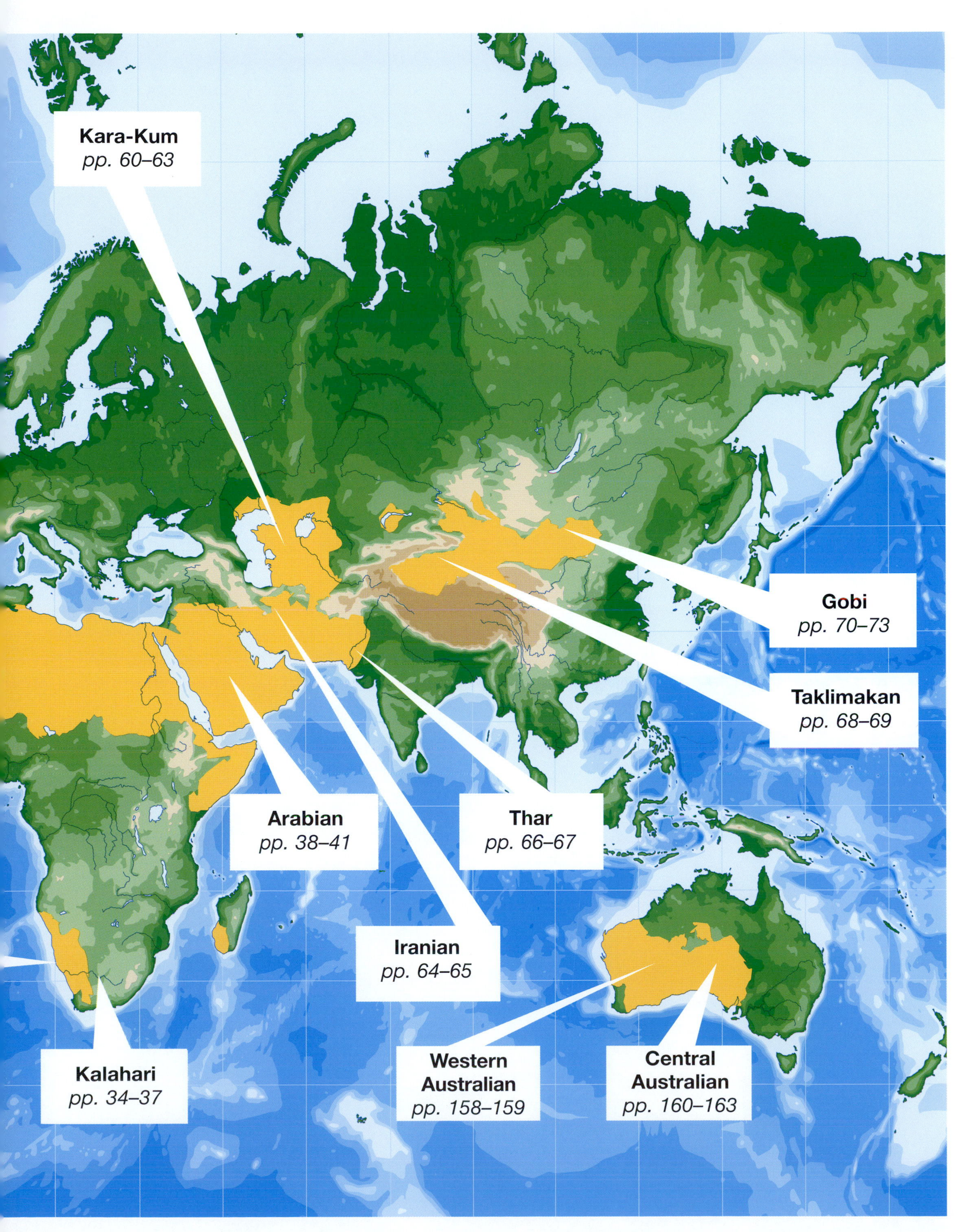
Kara-Kum
pp. 60–63
Gobi
pp. 70–73
Taklimakan
pp. 68–69
Arabian
pp. 38–41
Thar
pp. 66–67
Iranian
pp. 64–65
Kalahari
pp. 34–37
Western Australian
pp. 158–159
Central Australian
pp. 160–163

# HOW DESERTS FORM

*The answer to the question "Why do deserts form?" seems obvious – sustained lack of rainfall – but the global and local climatic conditions that lead to such aridity are complex and an understanding of them helps explain such apparent anomalies as coastal deserts.*

Deserts are among some of the most alien, inhospitable landscapes on the planet. Some of their most striking features – the vast fields of rubble, austerely patterned dunes, dry or seasonal riverbeds, gleaming rinks of sun-baked salt, and the seeming near-absence of life might lead an observer to suspect they are the result of some great global catastrophe. In fact, just like any other biome, or major habitat, such as rain forest, tundra, and steppe, the world's deserts have evolved over millennia – the result of complex interactions between climate and geology.

In this chapter we look at the climatic conditions, such as global wind and ocean currents and continental rainfall patterns, that have shaped both the deserts of the prehistoric past and those of today. Later, in Chapter 8, we will see how future climate changes – some of then human-caused – might shape the deserts of tomorrow.

**A dry delta curls through eroded hills in California's Death Valley, in the rain-shadow of the Sierra Nevada, marking where a river ends in a channel of sediment.**

## Prehistoric Deserts

Deserts have not always been where they are today. They have grown and shrunk and shifted around the planet over millions of years – a natural consequence of the great changes wrought upon the earth through the geological ages. Continents have drifted around the globe, sea levels have risen and fallen, temperatures have fluctuated, and climate patterns have shifted. Doubtless these changes, once natural but increasingly affected by human activity, will also shape the deserts of the future.

Deserts have probably never been so extensive on earth as they were during the Permian period, the last phase of the Paleozoic era, some 290 to 245 million years ago. At this time all of the main landmasses – the continents – were butted up against each other to form one giant block of land, the supercontinent known to geologists as Pangaea. The global climate of Pangaea was in some ways uniform, without such variations in temperature from equator to poles as exist today, for example. In terms of rainfall, however, the continent was far from uniform. Winds picked up moisture as they blew over the earth's seas and dropped this as rain near the

| ERA (MILLIONS OF YEARS BEFORE PRESENT) | PERIOD (MILLIONS OF YEARS BEFORE PRESENT) | |
|---|---|---|
| PALEOZOIC (570–245) | Permian (290–245) | "Age of Deserts" begins on the supercontinent Pangaea |
| MESOZOIC (245–65) | Triassic (245–208) | Age of dinosaurs begins |
| | Jurassic (208–146) | Global climate becomes warm and moist<br>Deserts in decline |
| | Cretaceous (146–65) | Earliest flowering plants<br>Extinction of dinosaurs |
| CENOZOIC (65–0) | Tertiary (65–1.6) | Earliest large mammals<br>Most modern deserts begin to form (13 million B.C.) |
| | Quaternary (1.6–present) | Earliest humans<br>End of last Ice Age (*c.*8000 B.C.)<br>Irrigation of Mesopotamian Desert<br>Emergence of Sahara (*c.*4000–2000 B.C.)<br>Desertification of drylands bordering deserts |

Although arid conditions were widespread in the Paleozoic "Age of Deserts," most modern deserts were formed relatively late in the Cenozoic period.

coasts. As they blew on and reached the supercontinent's vast inland regions, however, they became dry as bone. Deprived of almost all precipitation, huge tracts of the interior of Pangaea, far from the sea, were harsh, barren scrub or near-empty desert.

Geologists have surmised the existence and location of prehistoric deserts from the evidence of rocks and fossils. Rocks have been forming, breaking down, and reconstituting almost since the earth came into existence some 4,600 million years ago. Different types of rocks reflect the conditions of their formation. For example, limestones such as chalk are laid down on the beds of great seas, while coals originated as the lush, part-decomposed plants that thrived in ancient swamps. Dark basalts were once vast flows of molten rock, or lava, that oozed up from the depths of the earth and slowly cooled and solidified. The characteristic rocks that indicate the existence of prehistoric deserts are sandstones, which formed as grains of sand became buried, compacted, and "glued" together. (By a curious process of geological reversal, the sands of the "classic" modern desert usually result from the erosion of these same ancient sandstones; *see* p. 52.) Some sandstones are formed in shallow seas, but their detailed makeup differs sharply from that of dry-land desert sandstones.

During the Triassic period (245–208 million years ago) – the first phase of the Mesozoic era that followed the Paleozoic – many "red-rock" deposits

Betraying their origins in the compacted grains of ancient deserts, sandstone cliffs line the channel of Wadi Rum in the Israeli desert.

**This fossilized head of the birdlike dinosaur *Coelurosaur* was discovered in arid New Mexico, North America. Like specialized arid-land creatures of today, dinosaurs flourished in the relatively dry interior of Triassic Pangaea.**

were laid down. These widespread features of arid conditions are found, for example, in Australia's desert interior. The red rocks include sandstones, siltstones, and shales that have been colored red by the oxidation of one of their chief iron-containing minerals, hematite (ferric oxide).

### The evidence of fossils

The difference between the sea- and land-formed sandstones is made still clearer by the types of fossils that these rocks contain. While shallow-water sandstones contain the remains of fish, shell-fish, and similar sea life, embedded within desert sandstones are the fossilized bones, teeth, claws, and other parts of land-dwelling animals. Many of these were reptiles, which, as we shall see in Chapter 4, are peculiarly well-adapted to life in arid conditions. About halfway through the Triassic period the typically large reptiles known as the dinosaurs appeared (*see panel, left*). At or around the same time, the first, small, shrewlike mammals also evolved. Fossil evidence also shows that the plants that flourished in Pangaea were well-adapted to its arid conditions. Ginkgoes, seed ferns, cycads, and, increasingly, conifers all flourished.

The evidence of both rocks and the fossils that they contain has suggested to paleontologists the existence of a dry, rocky countryside with scattered patches of vegetation where reptiles, insects, and scorpions scratched a living. It was this general landscape that dominated the massive interior of the supercontinent, Pangaea.

Toward the end of the Triassic period, Pangaea began to break apart into smaller blocks, and the world's oceans extended inlets and arms deep into the gaps. As the newly formed continents drifted apart and the interiors of the landmasses grew nearer the sea, moisture-laden winds were able reach inland areas. By the start of the next great time span, the Jurassic period (208–146 million years ago), the climate on land had become warm and moist. Greenery spread rapidly, and the great "Age of Deserts" drew to a close.

### The formation of modern deserts

The majority of modern deserts began to take shape around 13 million years ago, while the distribution of deserts we see today seems to have been established by about three million years ago. However, many dry regions have shifted and fluctuated in size since then and continue to do so (*see* Chapter 8).

As we saw in the Introduction (*see* pp. 7–9), deserts are characterized by very low rainfall and other types of precipitation. However, this feature is

## DINOSAURS OF THE DESERTS

Arid, rocky regions with their lack of soil and plant cover are ideal for fossil-hunting. Remains of many dinosaurs, prehistoric mammals, and other long-gone creatures are regularly discovered in modern deserts such as the Gobi, the Patagonian Desert, and the Kalahari. A major expedition of the 1920s visited the Gobi in search of remains of our human ancestors. However, most of the surface rocks there are far too ancient, being formed long before even our most distant apelike ancestors existed on earth. What the paleontologists did find were the fossils of many dinosaurs, including the pack-hunting *Velociraptor*, which walked upright on its long rear legs; the four-legged, pig-sized horned dinosaur *Protoceratops*, along with fossil evidence of its nests and eggs; and the low, lizard-shaped *Psittacosaurus*, with its parrotlike beak.

In recent years many astounding fossil discoveries have been made in the dry lands of Argentina, including one of the earliest dinosaurs, *Herrerasaurus*, which had long hind legs and sharp teeth; and one of the biggest, perhaps weighing 100 tonnes (98 t.), *Argentinosaurus*.

not caused in the same way in every desert region but is the result of a complex combination of factors. This very complexity of causality in their formation explains what might appear to be as the anomalous or surprising location of some of the world's great deserts. Close to both the Arabian Sea and the great Indus River, for example, the existence of the Thar, or Great Indian, Desert, may seem strange. A better understanding of climate and, specifically, of why aridity prevails in certain regions – on both global and local levels – helps unlock such apparent mysteries.

As we shall see, in the majority of instances the key factor in desert formation is latitude – the position up or down the globe, north or south from equator to pole – with its concomitant effects on levels of rainfall. In other deserts, however, other elements are decisive, such as their distance from the sea or the presence of nearby mountain ranges. In still others, the balance of factors is more complex – a subtle amalgam of various contributing factors. In general, however, it is possible to group deserts into one of three major categories – subtropical high-pressure deserts, rain-shadow deserts, and continental deserts.

## THE CORIOLIS EFFECT

The surface of the earth moves fastest at the equator, while at more northerly or southerly latitudes, the surface speed becomes slower. This differential speed of rotation is the basis of the Coriolis effect, or force, by which an object moving due north or south from the equator retains more of its original eastward speed than the surface rotating below it.

This means that an object moving north or south is also deflected east. In a bathtub plug hole the Coriolis effect has the mundane result of making the water swirl in a spiral. In the atmosphere, air heated by the sun at the central tropics rises and moves north or south but also maintains some of its eastward speed, too. As this air cools, sinks, and returns to the central tropics, the reverse happens and it is deflected west. These create the northeast and southeast trade winds, as shown in Figure 1 below.

The planetary wind system is important for the location of some of the world's great deserts. For example, subtropical high-pressure winds blowing eastward from the Pacific strike the mountain ranges of the American continent and produce the rain-shadow deserts in North and South America.

Figure 1

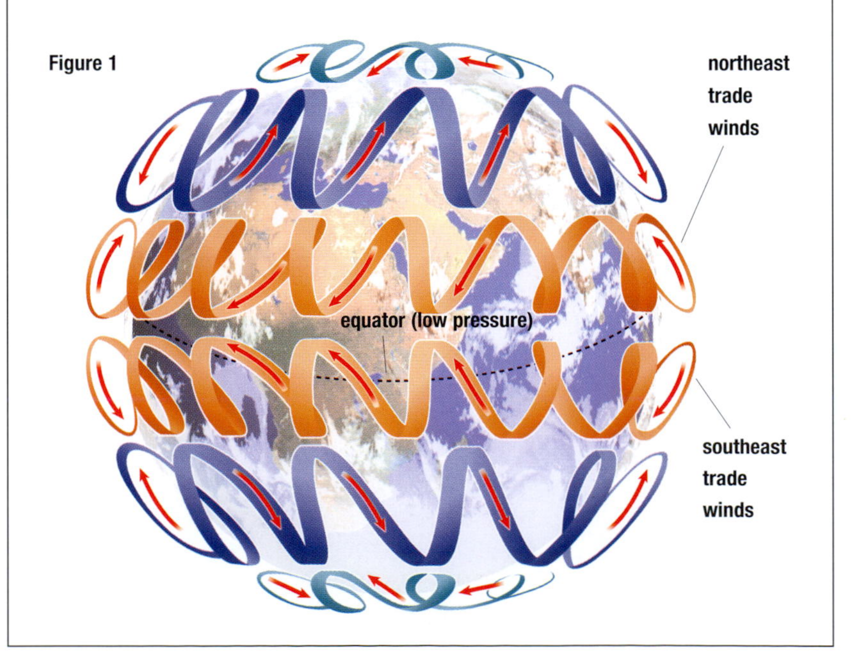


### Subtropical High-pressure Deserts

A large number of the world's deserts are found in regions immediately north of the tropic of Cancer (the parallel of latitude about 23½° north of the equator) or immediately south of the tropic of Capricorn (the parallel of latitude about 23½° south of the equator) – that is, in the so-called subtropics. The reasons for this arise from patterns of air and water movement across the earth, which are themselves the result of complex interactions between global phenomena, such as the earth's 24-hour, west–east rotation and solar energy.

#### Global wind patterns

In part, global wind patterns are caused by the way in which the sun's rays warm the earth. The part of the earth that receives most of the sun's heat is the tropics – the region about the equator that lies between the tropics of Cancer and Capricorn. In the tropics the sun's rays hit the earth almost at right angles, concentrating their heat energy on the smallest surface area. They also pass through the least depth of atmosphere, with minimal scattering and spreading, before they reach the surface. Farther north and south from the equator the sun's rays approach the earth at a slanting angle and their heat energy covers a correspondingly larger area. They also have to pass through a much greater depth of atmosphere, causing their heating effects to be spread and dissipated.

The greater heat at and near the equator means that the air there becomes hot and rises, allowing cooler air to flow in. The rising hot air might be expected to move away, due north and south. That this does not quite happen is due to the Coriolis effect, or force (*see panel*), named after the French civil engineer Gaspard Coriolis, who first noted the phenomenon in the 19th century. Under its influence the hot equatorial winds blowing north and south are deflected from west to east, while cooler winds drawn back to the equator are deflected from east to west. These masses of air are known as the trade winds, which blow steadily for much of the year from the northeast north of the equator and from the southeast to its south, roughly between latitudes 0 and 30°. (The term "trade" used to describe the winds derives from an obsolete meaning of the word – "in a regular course or direction" – but also reflects the winds' importance for merchant shipping.)

The warm, rain-drenched central tropics near the equator are home to the world's rain forests, such as those found in Amazonia. Many of the great deserts, by contrast, are found in the subtropics, which, at a greater distance from the equator, receive little rain.

## GLOBAL OCEAN CURRENTS

The planetary wind system plays an important role in the creation of the oceans' surface currents. In general, these currents follow the enormous circulation loops around the oceans, delivering a huge supply of heat from the equator to high latitudes. Most of the currents' heat is lost along the western boundaries of oceans, so that by the time they make their return journey along the eastern boundaries they are cold.

The coastal deserts of Africa and South America owe their existence to this phenomenon. For example, the Namib Desert (*see illus.*, p. 18) on the western coast of southern Africa is washed by the cold Benguela current. Cold air holds very little moisture, so the Namib receives little rainfall.

Occasionally the circulation of the oceans' surface currents is reversed. El Niño, which occurs every five to eight years, reverses the usual pattern of currents in the South Pacific. The results include severe drought in Australia and heavy rains and floods in the western countries of South America.

### Wet tropics, dry subtropics

The sun's heat at the tropics not only warms air, it also evaporates ocean water into water vapor that disperses into the air. The rising moisture-laden warm air rapidly expands and cools as its pressure reduces (since atmospheric pressure is highest at the earth's surface and decreases with height), and its moisture condenses back into water, falling as rain that is largely confined to a belt about 10° north and 5 to 10° south of the equator. This is why much of this belt, the central tropics, is extremely moist and covered with dense, lush vegetation.

The now almost moistureless air continues to gain height and flow northeast or southeast, pushed by more hot, moist air rising at the central tropics. Gradually it moves beyond about 20 to 25° north and south of the equator, into the subtropics, and becomes cool enough to descend. As it does so, its pressure rises, reheating the air, just as the air squeezed in a bicycle pump becomes hot. This is the warm, dry, high-pressure air found about 25 to 30° north and south of the equator. It is also the air that helps create most of the world's deserts.

### Circulation of air

Most of the dry, warm air that descends at 25 to 30° north or south loops back to move at surface level in the reverse direction of its outward journey. Gradually it is drawn back toward the tropics, is

**The Namib Desert, shown here in a satellite image, is maintained by the cold, dry Benguela current.**

recharged with heat and moisture, and so continues its movement. The end result is a cycle of air moving from the surface at the tropics up into the atmosphere, and then either northeast or southeast, sinking back to the surface at 25 to 30° north or south; this cooler air then returns toward the tropics from the northeast or southeast in the form of the trade winds. This air circulation forms corkscrewlike spirals north and south of the equator, known to climatologists as Hadley Cells.

The Hadley Cells are not self-contained. Some of the warm air descending at 25 to 30° north or south flows, not toward the equator again, but away from it, toward the middle latitudes. There it mixes with cold air coming from the polar regions, creating "battlefields" between warm and cold known as fronts. The fronts provide many temperate regions with their changeable weather.

### High-pressure deserts

We have seen how the interaction of atmosphere, winds, and ocean currents (*see panel*, p. 17) sets up belts of high atmospheric pressure at subtropical latitudes of about 25 to 30° north and south of the equator, and that most of the world's deserts are found in or next to these belts. Consequently it is can be said that there are no true deserts within 10° north or south of the equator, although there are some arid or semidesert regions, such as those that occur on the Horn of Africa. Likewise, there are no major arid areas beyond about 45° north or south, with the notable exception of the polar deserts (*see* pp. 22–23 *and* 164–167).

Deserts that owe their formation chiefly to these 25 to 30° north/south high-pressure zones are often known as high-pressure, or subtropical, climate deserts. Other deserts are sited in or near the 25 to 30° north/south zone but are maintained chiefly by additional factors, which are discussed below.

In the northern hemisphere, the two main examples of high-pressure climate deserts are the great Sahara (*see* pp. 26–31) and its eastern neighbor, the Arabian Desert complex (*see* pp. 38–41). Both these deserts receive air that originates from the moist tropical zone to the south but which has already given up its moisture. The center of the Sahara is made even drier because of its distance from the sea – the continental desert effect discussed below.

In the southern hemisphere the persistent high-pressure atmospheric features described above produce deserts at latitudes of around 25° south. These southern high-pressure climate deserts include the Sechura (*see* pp. 122–123) and Atacama (*see* pp. 124–127) deserts in South America, the Namib (*see* pp. 32–33) and Kalahari (*see* pp. 34–37) in Africa, and most of the deserts in Australia (*see* pp. 158–163).

### Rain-shadow Deserts

In some instances, the rain-shadow effect is the primary cause of desert formation. In others, it intensifies or confirms the arid conditions already set in place by other factors, or it may help to delineate the extent of desert areas.

The mechanics of the rain-shadow effect are shown in Figure 2 (*below*). Air gathers moisture from the sea in the form of water vapor and, as it moves inland, blows steadily against a mountain range. The windward slopes of the mountains –

**Figure 2 Rain-shadow deserts such as the Mojave and Great Basin deserts of the American Southwest form on the leeward side of mountains.**

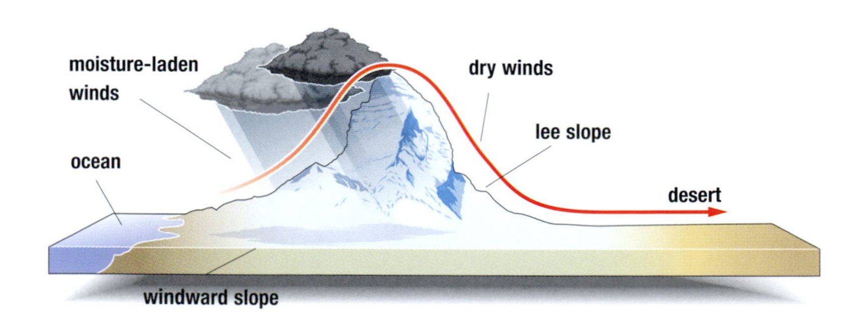

those facing the oncoming winds – make the air rise (a phenomenon known as orographic lifting; *oros* is the Greek word for "mountain"). Because atmospheric temperature falls with increasing height – by an average of some 6°C (11°F) for every 1,000 meters (3,280 ft.) – the air mass cools as it rises. And because cooler air can hold less moisture as water vapor compared to warm air, water vapor condenses into droplets that fall as rain on the windward side of the slopes, turning to snow with increasing altitude.

The air now moves on, over the peaks of the mountains and down the leeward slopes – those lying away from the prevailing winds. By now it has lost most of its moisture content, so that little or no rain falls, creating a dry region on the leeward side of the mountains. This area is not only drier but much less cloudy than the other side of the mountains. Without cloud cover, the warming effect of the sun's rays is maximized, and the little water that does fall is lost through evaporation. Thus, in the same way that high peaks block light coming from the sun and cast deep shadows on the opposite side, the peaks also block moist air and rain from the leeward side, hence the term rain-shadow.

The rain-shadow effect is most extreme when the mountains are tall and lie perpendicular to the prevailing winds, and when these winds shift only rarely from their regular direction. The effect can extend for hundreds of kilometers beyond the mountain range before other air masses are able to dilute it.

### North American rain-shadow deserts

Two deserts where the rain-shadow effect predominates are the Mojave (*see* pp. 118–121) and Patagonian (*see* pp. 128–129) deserts of America. The Mojave in California lies in the lee of the Sierra Nevada, a mountain range to its west. The range runs roughly north–south for more than 600 kilometers (370 mi.), with an average width of some 100 kilometers (62 mi.), and includes the highest mountain in the contiguous United States, Mount Whitney at 4,418 meters (14,495 ft.).

Winds blowing east from the Pacific Ocean shed their rainfall over the first line of hills, the Coastal Ranges of California, and then against the wind-

**The rain-shadow effect caused by the majestic Himalaya mountain system helps create some of the world's largest and most inhospitable deserts – the Gobi and the Taklimakan of Central Asia.**

## VALLEY OF THE DEAD

Barely 100 kilometers (62 mi.) from the highest peak in the contiguous United States, Mount Whitney in the Sierra Nevada, lies the lowest point in the western hemisphere in Death Valley. In this northern arm of the Mojave Desert, the valley floor descends to 86 meters (282 ft.) below sea level. The valley may not look large on a map – it is about 220 kilometers (136 mi.) long and between 8 and 25 kilometers (5 and 17 mi.) wide – but its searing daytime temperatures, regularly exceeding 50°C (122°F), can make it a dangerous place. Its dry, harsh climate results partly from a local rain-shadow effect of the Panamint Mountains to the west, accentuating the major rain-shadow from the Sierra Nevada. The valley was named from its effect on gold-seekers and settlers, many of whom died as they attempted to cross it to reach California in the mid-19th century. Even today, tourists and trekkers regularly become stranded in the valley, and fatalities still occur. This image from the Landsat-4 satellite shows the valley as a largely white area – the color is created by mineral salts – among barren highlands, colored brown.

ward wall of the Sierra Nevada. By the time the air reaches the Mojave, it is bone dry and the land resembles a moonscape, with salt flats, rocky outcrops, and sand-covered plains. The land surface of the desert was once the bed of the Pacific Ocean, until volcanic activity and mountain-building created the Sierra Nevada and Coastal Ranges and cut it off from the sea. A northern spur of the Mojave is the infamous Death Valley, the landscape of which reflects one of the most extreme climates found on the planet (*see panel above*).

### South American rain-shadow deserts

The Patagonian Desert in South America lies in the rain-shadow of the mighty Andes Mountains that run parallel to the western edge of the continent. Prevailing winds in this temperate area, some 40 to 50° south, are from the northwest and gradually lose their moisture content over the high Andean peaks, many of which approach 4,000 meters (13,000 ft.) in altitude. Cool, dry winds – known as *pamperos* – sweep down the western Andes foothills at an altitude of about 1,500 meters (4,900 ft.), and across the Patagonian plains. Their moisture evaporated, they carry away any rain that has fallen, thereby accentuating the drying effect of the Andean rain-shadow.

### Continental Deserts

Another factor in the formation and location of deserts is distance from sea. Air moving over seas and oceans collects water vapor evaporated from the surface by the sun's heat. As the moisture-laden air crosses the coast and blows over the land, it rises and cools, causing the water vapor to condense and fall as rain or snow.

The air gradually loses its moisture and moves on. Unless it crosses a major lake or river system, where more sun-powered evaporation can take place, the air's water vapor content is not replenished, and the farther it travels from the coast, the drier it becomes. Consequently, regions located deep inland, in the centers of continents, generally have dry climates; wherever winds may blow from, they have generally lost their moisture by the time they arrive. In some cases the dryness is severe enough to maintain deserts. Since these deserts are found toward the middle of major landmasses, they are termed continental deserts.

### Asian continental deserts

The greatest collection of continental deserts is found in Asia – unsurprisingly, given its vast land area of some 33,391,162 square kilometers (17,139,445 sq. mi.). The Kara-Kum (*see* pp.

60–63), Taklimakan (*see* pp. 68–69), and Gobi (*see* pp. 70–73) deserts have all formed mainly owing to the continental effect. The formidable Gobi, for example, is more than 2,000 kilometers (1,240 mi.) from any major body of water, while the virtually mountain-locked Taklimakan to its west is at a similar distance from the sea. The Kara-Kum, it is true, lies close to the Caspian Sea, but this relatively small, and shrinking, body of water is not sufficient to compensate for the desert's remoteness from oceanic shores and moisture-bearing winds.

These Asian deserts also experience rain-shadow effects from the world's largest, tallest mountain range, the Himalayas, which lies to the south and southwest. In this temperate zone, regular winds coming from the southwest drop heavy rain and thick snow on the Indian subcontinent and the windward sides of the peaks. To the north of the Himalayas, however, the air has run out of moisture. The Taklimakan is "rain-shaded" on the west and north, too – by the lofty Hindu Kush and Tian Shan mountain ranges respectively.

### Australian continental deserts

The continental effect also contributes to maintaining the great inland deserts of Australia (*see* pp. 158–163). Indeed, this "desert continent" has contributions from all three major desert-creating processes. The southern high-pressure climate belt, centred on the 25° south latitude, runs across the landmass. The Gibson and Simpson deserts are mostly more than 1,000 kilometers (620 mi.) from any coastline, while the Great Dividing Range, which runs down Australia's east coast, has a rain-shadow effect due to prevailing easterly and southeasterly winds. This is another reason why the western half of Australia, where the Great Sandy, Gibson, and Great Victoria deserts are located, is drier than the eastern.

**The Gobi Desert of Central Asia is an important example of a continental desert. Covering much of the land-locked country of Mongolia and the Chinese province of Inner Mongolia, it is thousands of kilometers from the nearest ocean.**

# POLAR DESERTS

The term "desert" was once confined to hot, arid regions of the tropics and subtropics. However, the central place given to the absence or sparsity of vegetation in many definitions of the term has led to the development of the concept of the "cold desert" – a terrain in which low temperatures and physiological drought drastically inhibit plant growth. Such conditions are, of course, most typically found in the earth's polar regions – the Arctic and Antarctica – leading to the concept of the "polar desert."

True polar desert is surprisingly rare in the Arctic (*see* pp. 164–165). Most of the central north polar region comprises the Arctic Sea; the Arctic's land area is made up of the northernmost parts of the North American and Asian continents as well as of a large number of islands, notably Greenland, the Canadian Arctic Archipelago, and the Svalbard group. Lack of snowfall and relatively moderate temperatures, created especially where Arctic climates are tempered by maritime currents, also mean that less than two-fifths of this land is permanently covered with ice.

More typical of the Arctic is the low-lying scrubland that lies beyond the tree line – usually known as tundra, but in its American incarnation often known as the Barren Grounds. However, a few areas, such as the heavily glaciated inland of Greenland and its arid northern coast, resemble hot deserts in that plant life is minimal (*see* Chapter 3). However, as in the hot desert environment, some plants have evolved strategies to survive not only the bitter cold but also the long winter nights and long summer days. Many Arctic plants reproduce asexually, and many have a swift life cycle – growing, blooming, and seeding in a few weeks during the brief Arctic summer. Exposed rocks are home to numerous species of lichen, while even permanent ice supports algae.

Animals have adapted less well to Arctic conditions, perhaps because the relative swiftness of glaciation in the region meant that there was little time for adaption to take place. In the Arctic and Subarctic just a few species – such as the polar bear and fox, snowy owl, and gyrfalcon – tend to exist in relatively large numbers.

In contrast to the Arctic, Antarctica is a continent all of its own and is almost entirely covered with ice; indeed, the Antarctic ice sheet accounts for some 90 percent of the

**Adélie penguins (*Pygoscelis adeliae*) dot the landscape of the Adélie Coast of Antarctica. Animal life in the continent's White Desert is confined to its coastal regions, where the rich sea life provides food for several species of penguins and other birds.**

**Snow petrels (*Pagodroma nivea*) fly across the Antarctic. Some hardy birds of the region are known to have crossed the central polar region of the continent.**

earth's ice. This is the so-called White Desert, where winter temperatures can fall as low as -89.2°C (-128.6°F), far colder than even the Arctic. The continent is also subject to fierce blizzards, as cold winds sweep down from the interior highlands and sweep up the loose snow – just as hot desert winds whip up sands into blinding dust storms. Moreover, despite the presence of vast quantities of water in the form of ice, there is little precipitation in Antarctica – the equivalent of only about 50 millimeters (2 in.) rainfall.

Extreme cold, high winds and blizzards, and the scarcity of moisture mean that little plant or animal life can survive on Antarctica, although, by contrast, the surrounding seas are home to a rich variety of flora and fauna. Cold-adapted life tends to congregate around the more hospitable microclimates created where high altitudes combine with northerly, coastal latitudes to attract a higher degree of solar radiation. In total, there are only about 800 plant species in Antarctica, of which almost half are lichens, which are able to remain dormant during the long winter months. Animal life is still rarer. Apart from tiny arthropods such as mites and lice, the principal Antarctic animals are birds, most notably Adélie and emperor penguins, whose immense rookeries are crammed along the coastline. Other Antarctic species include petrels, albatrosses, and cormorants. With the exception of the great emperor penguin, all these birds migrate northward with the growth of the winter ice.

Hot deserts are remarkably stable environments. Their polar counterparts, by contrast, have proved much more susceptible to human interference. The development of whaling stations in the past and today's scientific stations and tourism have introduced many alien species of plants and even animals; on some subarctic islands rabbits and sheep have decimated local plant species. Of more concern is the threat of global warming and the melting of both Arctic and Antarctic glaciers, although as yet no clear environmental trends are discernible. Antarctica's probable great mineral wealth is a great temptation for the world's nations. At present, however, international treaties protect the White Desert from mineral exploitation, and there are plans afoot to make the continent a vast "world park."

KARA-KUM
IRANIAN
THAR
SAHARA
ARABIAN
INDIAN OCEAN
NAMIB
KALAHARI
SOUTH ATLANTIC

# African Deserts

*More than a third of Africa – the earth's second largest continent – is desert. Most of this comprises the vast Sahara, which sprawls across the north of the continent. Equally spectacular, though, are the deserts of the south – the coastal cold-water Namib and the inland Kalahari with its wealth of wildlife. Bridging Africa and Asia is the desert peninsula of Arabia, whose sands and plains are rich with oil.*

A camel caravan crosses the rolling sand dunes of the Ténéré Desert in eastern Niger. The Ténéré is one of the most remote and unspoiled parts of the Sahara.

# Sahara Desert

**1. Highest peak**
At 3,415 meters (11,205 ft.), Emi Koussi in the Tibesti Mountains of Chad is the highest point in the Sahara.

**2. Aïr Massif**
This mountainous area is a homeland of the nomadic Tuareg (*see* pp. 133–134). The massif, with its distinctive sharp black peaks, is mined for its uranium deposits.

**3. Tombouctou**
This small town in Mali, otherwise known as Timbuktu, was once an important trade center on Saharan trade caravan routes, a center of Islamic learning, and the capital of the powerful Songhai Empire.

**4. Oil and gas**
The Sahara's richest oil wells are found at In Salah, Algeria.

**5. Flood warning**
In 1922 the town of Tamanrasset in Algeria was destroyed by flashflooding.

**6. Tademaït Plateau**
This windy desert plateau has some of the most spectacular examples of desert varnish – a dark, hard glaze that forms on many of the Sahara's exposed rocks. The phenomenon is caused by coatings of iron oxide or manganese oxide, which are drawn to the surface by capillary action.

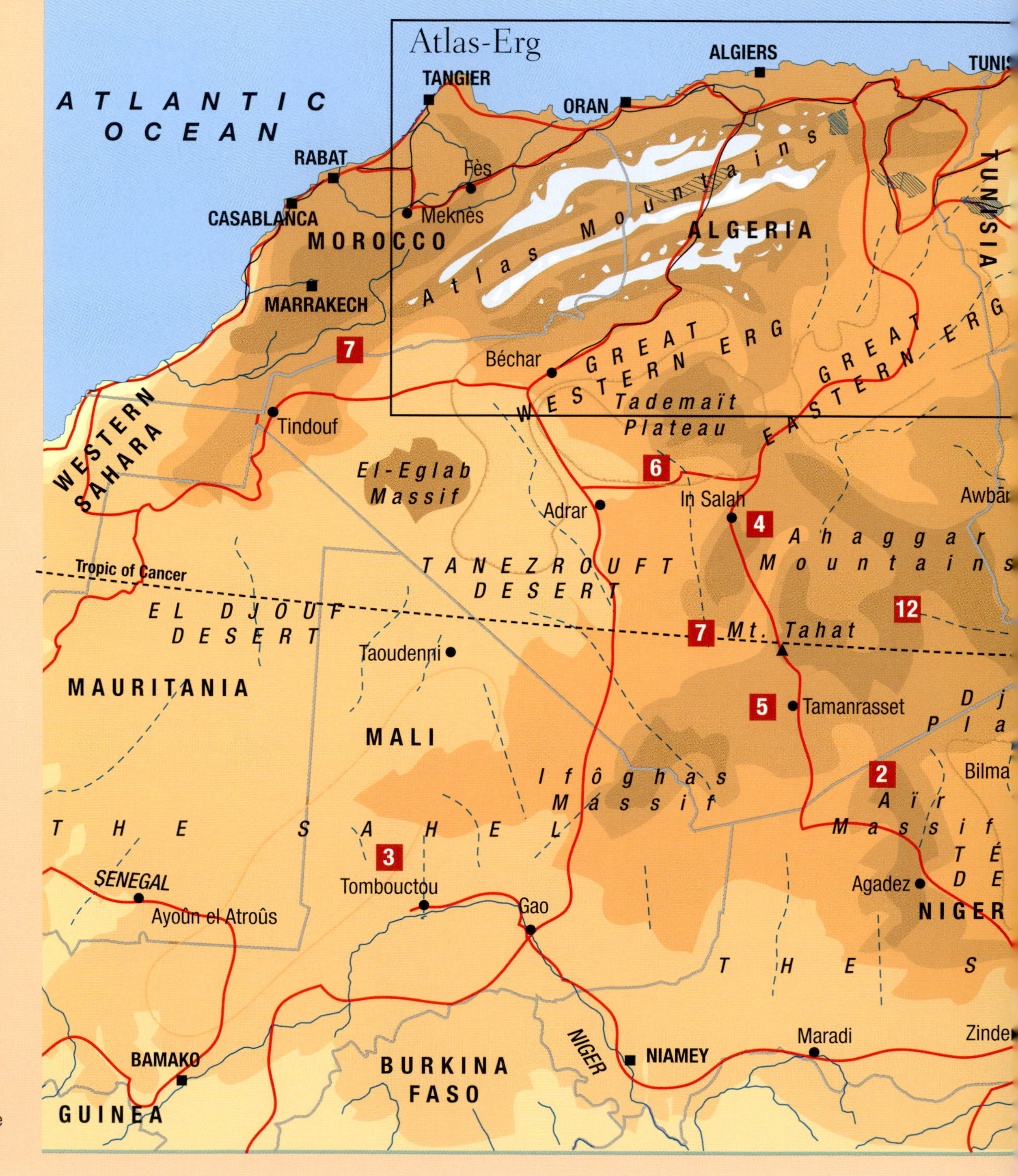

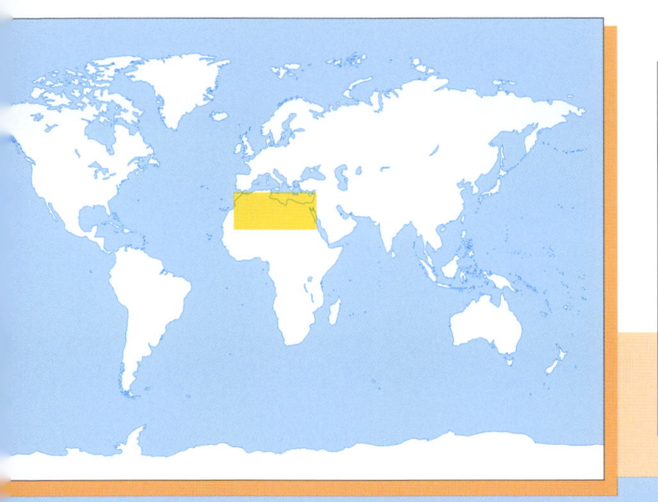

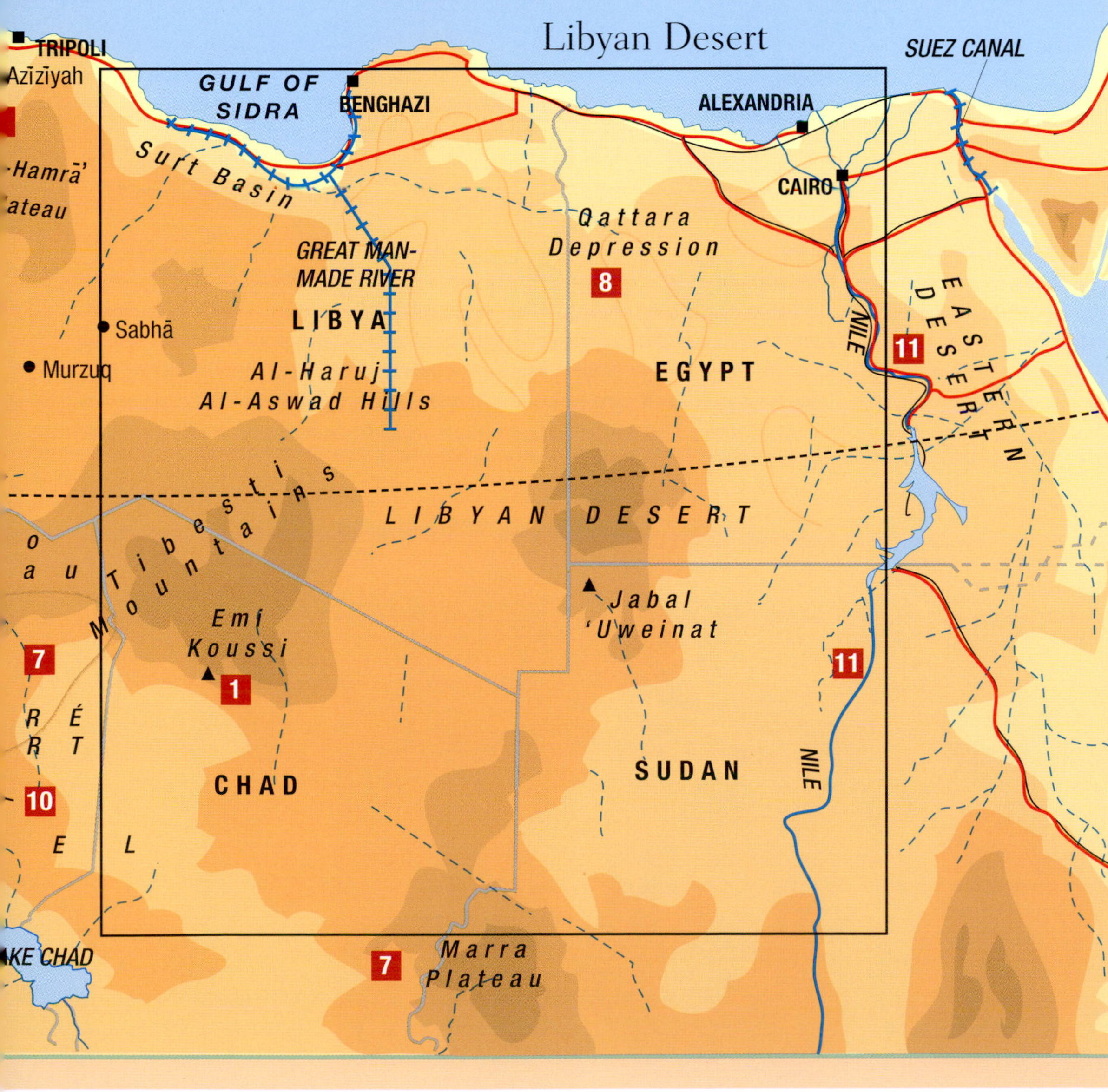


## 7. Wadis

Crisscrossing the Sahara are numerous seasonal streams, known as wadis after their Arabic name. Many flow from mountainous areas, such as the Tibesti and Ahaggar ranges, after heavy rainfalls.

## 8. Lowest point

The lowest point in the Sahara is the Qattara Depression in Egypt, which is 134 meters (440 ft.) below sea level.

## 9. Hot spot

The highest temperature ever recorded in the Sahara was 58°C (136°F) at Al-'Azīzīyah in Libya.

## 10. Ténéré Desert

This remote desert has some of the most beautiful sand dunes in the Sahara. It is famous for the Arbre du Ténéré, a tree (now replaced with a metal pole) reputed to be the most isolated on earth, as well as its dinosaur remains.

## 11. Nile River

Famously the longest watercourse in the world, the Nile is in its desert courses termed an exotic river because its sources lie far beyond the Sahara's borders.

## 12. Ahaggar Mountains

This large plateau region sprawls across the center of the Sahara, in Algeria. Its highest peak is Mount Tahat (3,000 m.; 9,842 ft.)

## Fact File

- The Sahara is the world's largest desert, covering some 9,065,000 square kilometers (3,500,000 sq. mi.) – an area only slightly smaller than the United States. The desert accounts for some 8 percent of the earth's land area.

- The Sahara includes parts of 11 countries: Algeria, Chad, Egypt, Libya, Mali, Mauritania, Morocco, Niger, Sudan, Tunisia, and Western Sahara (occupied by Morocco).

- The Sahara's name derives from the Arabic *sahara*, meaning "deserts," and indeed its vast land area in fact comprises several deserts, including the Eastern (or Arabian), El Djouf, Libyan, Nubian, Tanezrouft, Ténéré, and Western deserts.

- In the Sahara are several pyramid-shaped dunes that tower more than 152 meters (500 ft.) high.

- Only around 2.5 million people live in the Sahara, giving a population density of less than one person per square mile (0.4 sq. km) – one of the lowest on earth. The population is mainly nomadic, with settlement limited to areas immediately around oases, and large parts of the desert completely empty.

- Dust from the Sahara is occasionally carried as far afield as the United Kingdom and Germany.

- The elevation of the Sahara ranges from the highest point, Emi Koussi (3,415 m.; 11,205 ft.) to Egypt's Qattara Depression, at 134 meters (440 ft.) below sea level.

# The Largest Desert

The Sahara Desert is the world's largest desert. Stretching across the northern third of the African continent, from the Atlantic Ocean to the Red Sea, it covers some 9,065,000 square kilometers (3,500,000 sq. mi.). Straddling the tropic of Cancer, the Sahara is a classic example of the high-pressure desert, maintained by the warm, dry air belts found about 25 to 30° north and south of the equator (*see* pp. 16–18). The desert's dryness also increases toward the interior, as the continental effect (*see* pp. 20–21) takes hold.

Although the Sahara is often thought of as having a single, homogeneous terrain, it is in fact extremely varied, with only a quarter of the Sahara made up of the "classic" desert covering of sand dunes and seas known as erg. Other characteristic terrain types are the desolate gravel-strewn plains known as regs; hamadas, which are flat, bare-rock plateaus; and mountainous regions. Another characteristic feature of the Sahara are the numerous oases (*see* pp. 50–51) and wadis – seasonal streams – that are so vital to the survival of the desert's human, plant, and animal populations. The Saharan climate is similarly diverse, with large temperature variations across its extent. However, rainfall is scant and erratic almost everywhere and often comes in the form of sudden localized storms. Strong, unpredictable winds such as the khamsin and sirocco can create hazards such as dust storms and are responsible for some remarkable features such as desert varnish and sand dunes.

The desert has been inhabited by nomadic peoples such as the Tuareg for more than 7,000 years. Under Arab rule numerous trans-Saharan trade routes developed (*see map*, p. 134), carrying salt to sub-Saharan Africa and returning with gold and slaves. However, in those areas where water was more plentiful, notably along the Nile and Niger rivers and on the Mediterranean seaboard, settled civilizations were able to develop. In modern times the Saharan interior remains only sparsely inhabited.

**The southern part of the Atlas Mountains is subject to the influence of the Sahara Desert – strong winds have created a rugged, heavily eroded bare-rock landscape.**

# Atlas-Erg

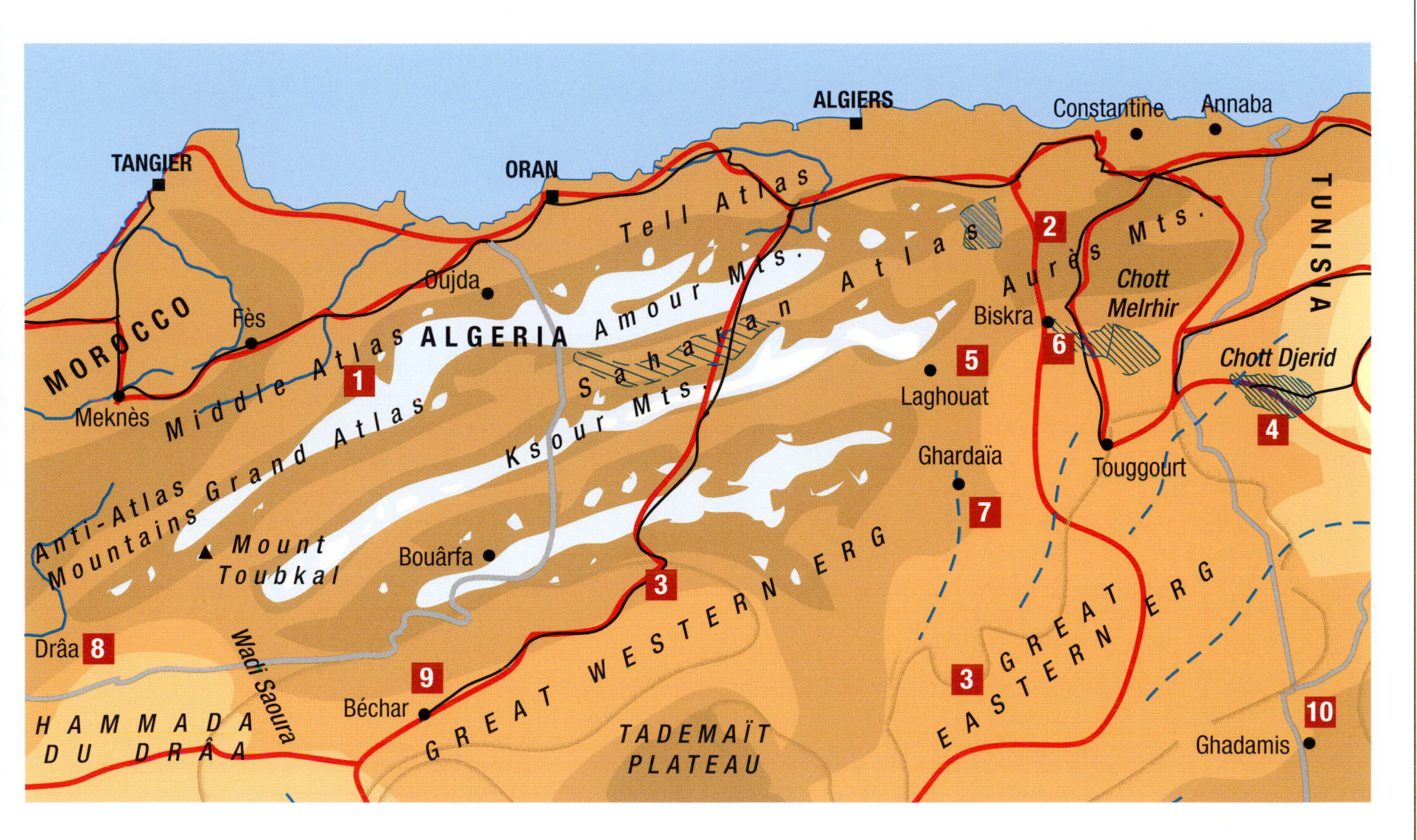


### 1. Atlas Mountains

These rugged mountains, comprising a series of roughly parallel ranges, mark the northwestern border of the Sahara Desert. Receiving high levels of rainfall, they include many fertile valleys. At 4,167 meters (13,671 ft.), the highest peak of the Atlas is Mount Toubkal. There are rich mineral deposits, including gold, silver, lead, and petroleum.

### 2. Aurès Range

The remotest part of the Atlas Mountains is home to the indigenous Shawia people, who are seminomadic farmers.

### 3. Ergs

An erg is an extensive region of sand dunes, or sand sea. The Great Eastern and Great Western ergs of the northwest Sahara are among the desert's most inhospitable landscapes.

### 4. Chott Djerid

A feature of the arid landscape to the south of the Atlas ranges are seasonal saline lakes called chotts (from the French form of the Arabic shatt). In ancient times one of the largest, Chott Djerid in Tunisia, was reputed to be the birthplace of the classical Greek goddess Athena.

### 5. Laghouat

This oasis town of about 70,000 people on the northern edge of the Sahara Desert is an important administrative and military center for the region and is well known for rug and tapestry weaving. In the surrounding region are rich natural-gas deposits.

### 6. Biskra

This oasis town was in ancient times the Roman military base of Vescera. After 1844 it served as a French base for military operations in southern Algeria. The surrounding oasis is an important producer of dates.

### 7. Ghardaïa

The chief town of the M'zab, a stony, barren valley of the northern Sahara, Ghardaïa is a center of date production and of the manufacture of rugs and cloth. The town was founded in the 11th century by members of the Muslim Kharijite sect fleeing persecution by the orthodox Muslims from the north.

### 8. Rivers

Numerous seasonal rivers, or wadis, flow down from the Atlas into the desert, but quickly peter out in the hot sun. Only a small number of rivers usually flow throughout the year here. These include the Drâa, Ziz, and Guir.

### 9. Béchar

Formerly called Colomb, this oasis town was a post of the French Foreign Legion. Nearby lie rich coal deposits.

### 10. Ghadamis

Often called the "Jewel of the Desert," this Libyan oasis settlement stands on the eastern edge of the Great Eastern Erg, at the junction of ancient caravan routes. It is unusual in having its palms and orchards within its walls, and for its rich Roman remains.

# Libyan Desert

**1. Highest peak**
At 1,934 meters (6,345 ft.), the highest point in the Libyan Desert is Jabel 'Uweinat in northwest Sudan, close to the border with Chad and Egypt.

**2. Siwa**
In ancient times this oasis was the seat of a famous oracle, sacred to Zeus Ammon. Ammonia was named for the god, near whose oracle the gas was first prepared.

**3. New Valley**
In 1999 Egypt launched the New Valley Project to irrigate thousands of square miles in the Western Desert, using the waters of the Nile.

**4. Al-Kufrah**
This group of five oases in eastern Libya covers some 18,130 square kilometers (7,000 sq. mi.).

**5. El Alamein**
This village in Egypt was the site of an important World War II battle in which the Allies halted the German advance on Alexandria and the strategically vital Suez Canal.

**6. The pyramids**
In the Western Desert near Giza are the Great Sphinx and three pyramids that date from pharaonic Egypt's IVth Dynasty (*c.* 26th century B.C.)

**7. Surt**
In former times this Mediterranean port was the starting point for many trans-Saharan caravan routes.

**8. Surt Basin**
Ever since oil was discovered here in 1956, this depression has been the greatest source of Libya's oil wealth.

**9. Pipelines**
Libya's crude oil is transported from the drilling rigs to the refineries by a network of surface pipelines. The first of these, linking the Zaltan field with the Marsa al-Burayqah refineries, was completed in 1961.

**10. Great Man-made River**
Opened in 1991, this $5-billion project exploits water reserves deep under the desert surface. More than 1,000 wells have been sunk and 3,500 kilometers (2,190 mi.) of primary pipeline laid to bring water to a region roughly the size of western Europe.

**11. Desert glass**
Libyan desert glass, found in the Great Sand Sea, is a natural glass composed of nearly pure silica. Some scientists suggest that the glass results from a meteorite impact on a silica-rich target.

## The Libyan Desert

The Libyan Desert is one of the least hospitable regions of the Sahara. The desert stretches from eastern Libya through southwestern Egypt, where it is known as the Western Desert, into northern Chad and northwestern Sudan. Much of the region comprises bare, rocky plateaus, interspersed with vast stony plains or sand seas.

The desert's wildlife and human life is concentrated around a few oases, including the Siwa, Baharīya, and Khārga oases in Egypt and the Al-Kufrah Oasis in Libya. Date-palms (*below*) make an important contribution to the oases' economies. Large-scale irrigation along the Great Man-made River in the west of the region has dramatically improved the fertility of the land.

Plant and animal life are restricted in diversity and number, although saltwort, spurge flax, goosefoot, wormwood, and asphodel grow in pockets in the semiarid zone toward the coast. Fennec foxes, hyenas, and adders inhabit oases and waterholes. Bird species include partridges, larks, hawks, and eagles.

The Libyan Desert is particularly important in human terms for its rich reserves of oil and natural gas, discovered in the 1950s, which are the mainstays of the Libyan economy. The richest reserves lie in the Surt Basin.

**Sand dunes dwarf an oasis in the Libyan Desert. Human habitation of this awesome landscape is confined almost entirely to a small number of such oases.**

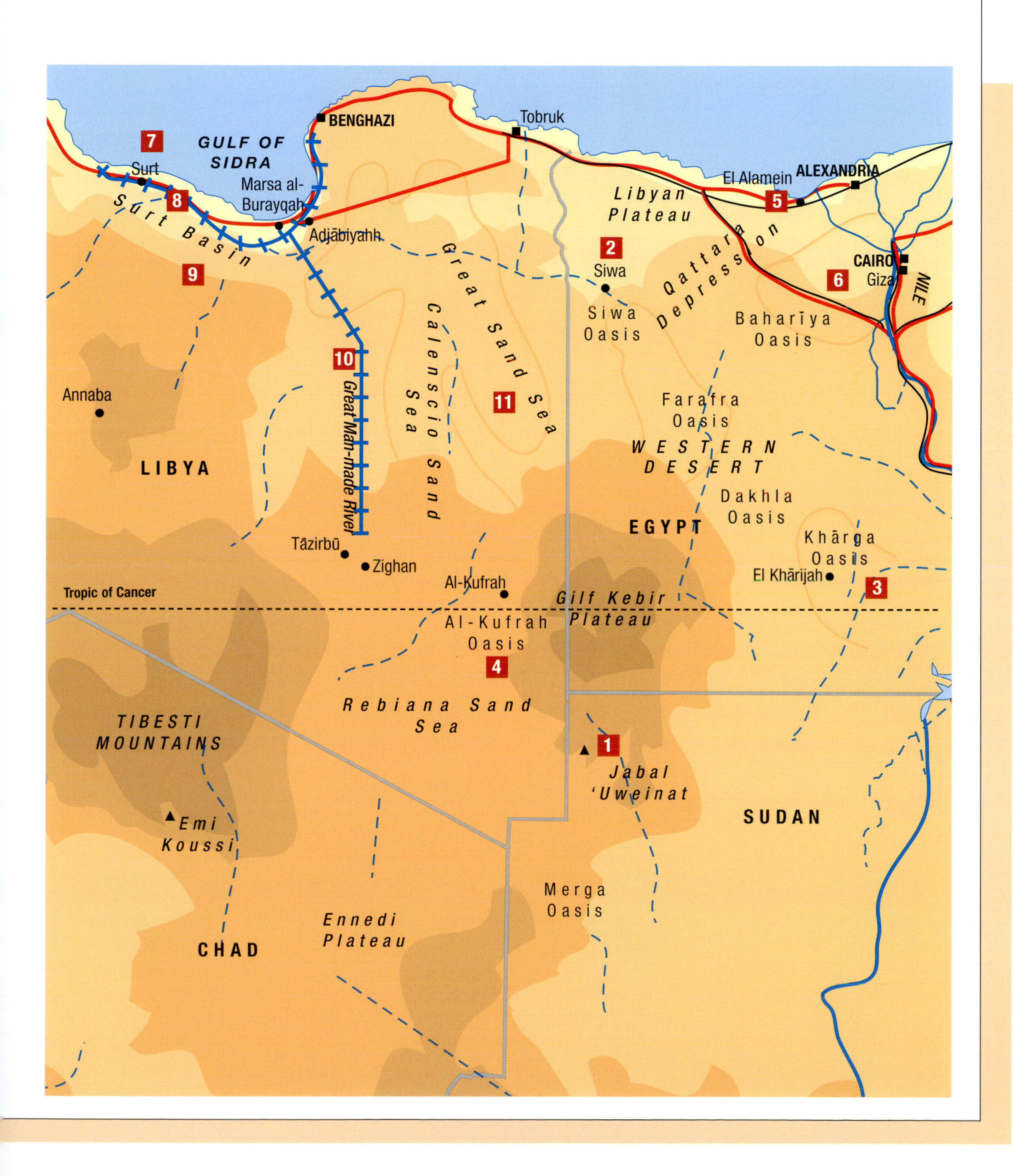

GULF OF SIDRA
BENGHAZI
Tobruk
Surt
Marsa al-Burayqah
Adjābiyahh
Surt Basin
El Alamein
ALEXANDRIA
Libyan Plateau
Qattara Depression
CAIRO
Giza
NILE
Siwa
Siwa Oasis
Great Sand Sea
Calenscio Sand Sea
Baharīya Oasis
Farafra Oasis
WESTERN DESERT
Great Man-made River
Annaba
LIBYA
EGYPT
Dakhla Oasis
Khārga Oasis
El Khārijah
Tāzirbū
Zighan
Al-Kufrah
Tropic of Cancer
Gilf Kebir Plateau
Al-Kufrah Oasis
Rebiana Sand Sea
TIBESTI MOUNTAINS
Jabal 'Uweinat
SUDAN
Emi Koussi
Merga Oasis
Ennedi Plateau
CHAD
1
2
3
4
5
6
7
8
9
10
11

# Namib Desert

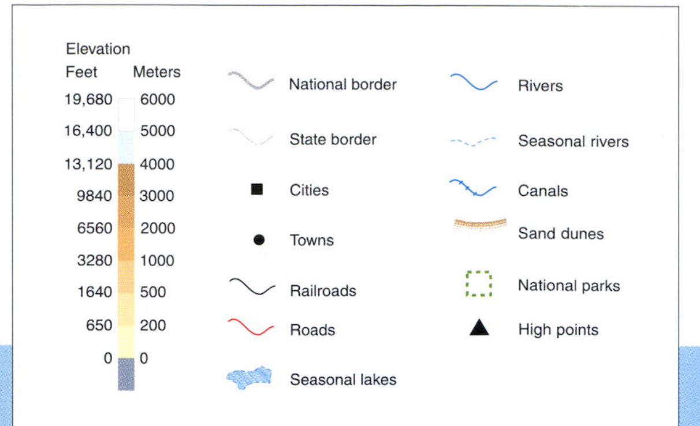

**1. Skeleton Coast**
This desolate region along the Namibian coast north of Swakopmund reputedly gets its name from the remains of the shipwrecked sailors who made landfall here only to die in the wastes of the Namib Desert. Today the area is a national park, which can be visited only as part of an organized group.

**2. Namib-Naukluft**
The Namib-Naukluft Park, located in the central Namib, is a massive desert and semidesert region, encompassing the diverse habitats of the Namib Desert Park, the Naukluft (formerly the Naukluft Mountain Zebra) Park, the dune fields of Sossusvlei, and the bird lagoon at Sandwich Bay.

**3. Sossusvlei**
This vast ephemeral pan in the Namib-Naukluft Park stands amid towering red sand dunes that stretch from the Koichab River in the south to the Kuiseb River in the north. The "sand mountains" at Sossusvlei rise to 300 meters (985 ft.) high.

**4. Etosha Park**
This large national park extends some 23,175 square kilometers (8,950 sq. mi.) over Namibia's central plateau, and is an important haven for some 93 mammal species and 340 bird species.

**5. Etosha Pan**
This flat saline desert, roughly 130 kilometers (80 mi.) long by 50 kilometers (30 mi.) wide, in the Etosha National Park is believed to have originated over 12 million years ago, as a shallow lake fed by the Cunene River. Climatic and tectonic changes have since lowered the water level so that the pan only holds water for a brief period each year, when it teems with flamingos and pelicans.

**6. Pipeline**
A 130-kilometer (80-mi.) pipeline carries water from the Koichab River to the coastal town of Lüderitz. The river itself has no source in the accepted sense but, like most of the watercourses of the Namib, terminates in sand dunes. The only rivers that flow constantly on the surface from spring to sea are the Cunene and the Orange.

**7. Windhoek**
In high summer (December to February), the capital of Namibia is so hot that many government departments move to the coastal resort of Swakopmund.

**8. Oranjemund**
The alluvial deposits at the mouth of the Orange River near the South African border are currently Nambia's main source of diamonds.

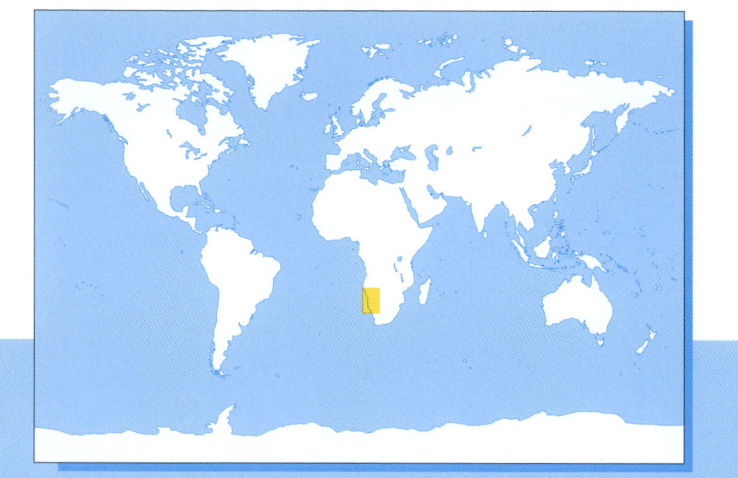

## A Coastal Desert

The long, narrow Namib Desert stretches along the coast of the southern African country of Namibia. Its semiarid fringes reach into Angola in the north and South Africa in the south, and blend with those of the Kalahari Desert to the east. The Namib is a cold-water coastal desert (*see* p. 17 *and illus.*, p. 18), its desolate terrain cooled and kept dry by the cold Benguela current, which runs northward offshore. The cold Atlantic waters also produce thick fogs, which drift inshore on sea breezes, as well as heavy dews, on which much of the desert's wildlife relies for moisture.

The Namib's landscape varies widely. Toward the south are towering orange, red, and apricot-colored dunes blown into razor-sharp ridges by the sand-shifting winds (*see illus.*, p. 54). To the north is the bleak Skeleton Coast, where the sands are more often than not swathed in thick fog. Inland are plains and flat-topped, steep-sided mountains known as inselbergs. Rivers crisscross the region, but are almost always dry. Toward the east are tablelands honeycombed with canyons and caves.

The Outer Namib – the strip of the desert closest to the sea – is almost completely barren, apart from a few succulents that are specially adapted to draw in precious moisture from the sea mists. Few mammals can live here, but numerous species of insects, such as the darkling beetle (*see* p. 97 *and illus.*, p. 88), and reptiles, including the desert sidewinding viper (*see illus.*, p. 91), thrive. In the sand dunes of the Inner Namib trees such as acacia grow along the larger river channels and mammals such as gemsbok are able to survive.

The Namib is largely unoccupied by humans, except for a few isolated coastal towns such as Swakopmund and Walvis Bay. In addition there are a few small groups of San (*see* pp. 138–139) and Khoi nomads – descendants of the country's first peoples. Vast areas of the Namib have been set aside for recreation and conservation.

**The red sand dunes of the Namib Desert southeast of Walvis Bay are divided from the Namib's rocky desert interior by the fertile canyon of the Kuiseb River.**

## Fact File

- The Namib stretches around 1,300 kilometers (800 mi.) from north to south and from 50 to 160 kilometers (30–100 mi.) inland.

- As a coastal cold-water desert, the Namib is among the world's coolest deserts, with inland summer temperatures averaging only around 31ºC (88ºF). Deserts on a similar latitude normally average over 38ºC (100ºF).

- The Namib derives its name from a word in the Nama language meaning "the place where there is nothing." Most of the Namib is totally uninhabited.

- Much of the south part of the Namib is covered in a vast expanse of sand, which tends to be redder inland, more yellow toward the coast. It gathers in great longitudinal, or seif, dunes that run from northwest to southeast, in accordance with the prevailing wind. The brilliant colors of the dunes are caused by slow iron oxidation; generally speaking, the brighter the color of the sand, the older it is.

- The north part of the desert, known as the Kaokoveld, is more rocky than the south. It has gravel plains, rocky outcrops, and scattered mountains, and in places has been carved into steep gorges by rapidly flowing seasonal streams.

- The Namib's sands originate mainly from the rivers that flow out of the eastern highlands, but which generally do not reach the sea, terminating in salt pans and alluvial deposits. Only the Orange and Cunene rivers run permanently on the surface.

- Some authorities consider the 300-meter (985-ft.) sand dunes at Sossusvlei in the Namib-Naukluft Park among the highest in the world. One of the very tallest is the much-photographed and rather prosaically named Dune 45.

- Many geologists believe that the Namib Desert is the oldest desert in the world, created more than 80 million years ago.

- One of the Namib Desert's most famous inhabitants is a plant. The welwitschia (*see illus.*, p. 84), found exclusively in the central part of the desert, can live for 1,000 years, or even longer, and has only two leaves. It is a succulent, gathering moisture from the fogs that blow inland from the sea.

- The average rainfall in the desert ranges from only 13 millimeters (0.5 in.) at the coast to around 5 centimeters (2 in.) at the Great Escarpment. In some years, however, no rain falls at all.

# Kalahari Desert

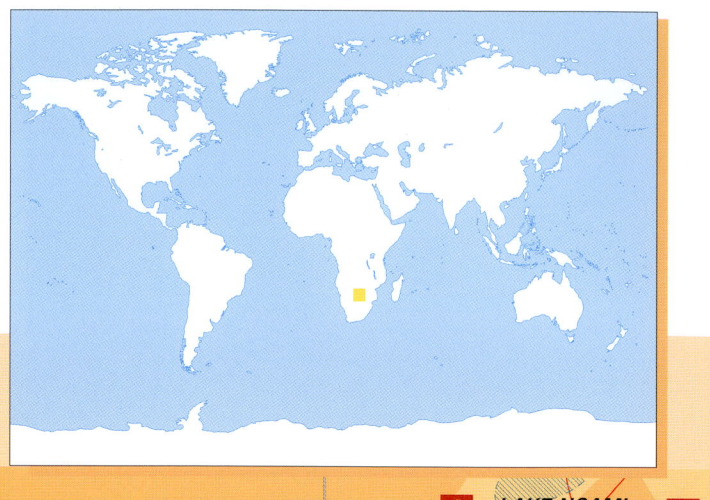

**1. Dry rivers**
The Kalahari holds hardly any permanent surface water. In the central and southern portions dry rivers, such as the Molopo, Okwa, and Nossob, traverse the terrain, providing invaluable wildlife habitats (*see illus.*, p. 105).

**2. Pans**
In the main expanses of the Kalahari what little surface water there is has become concentrated in ephemeral formations known as pans. The desert's few villages tend to cluster around the pans.

**3. Okavango Delta**
This vast inland delta (just off map) marks the northern limit of the Kalahari. While most of the world's rivers flow into the sea or a lake, the Okavango River is unusual in that, overwhelmed by the sandy aridity of the desert, it peters out into an inland delta. The delta irrigates the swamplands of northern Botswana, with their dense growths of reeds, papyrus, and pond lilies.

**4. Lake Ngami**
Following heavy rainfall in Angola, excess water in the Okavango Delta finds its way into this lake, which overflows into the Boteti River and ends up in the Makgadikgadi Pans. Lake Ngami was first recorded by the explorer David Livingstone in 1849.

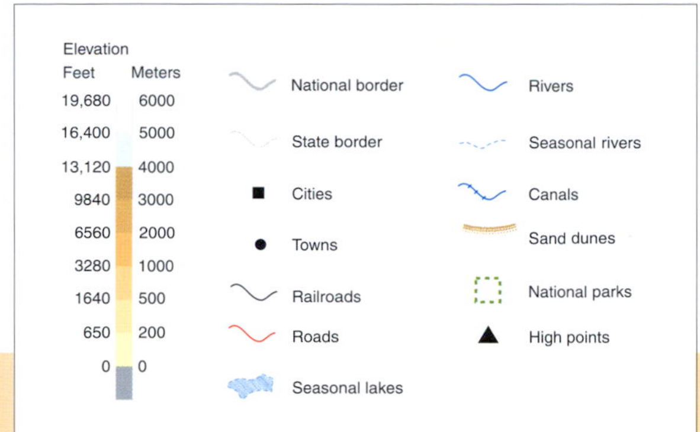


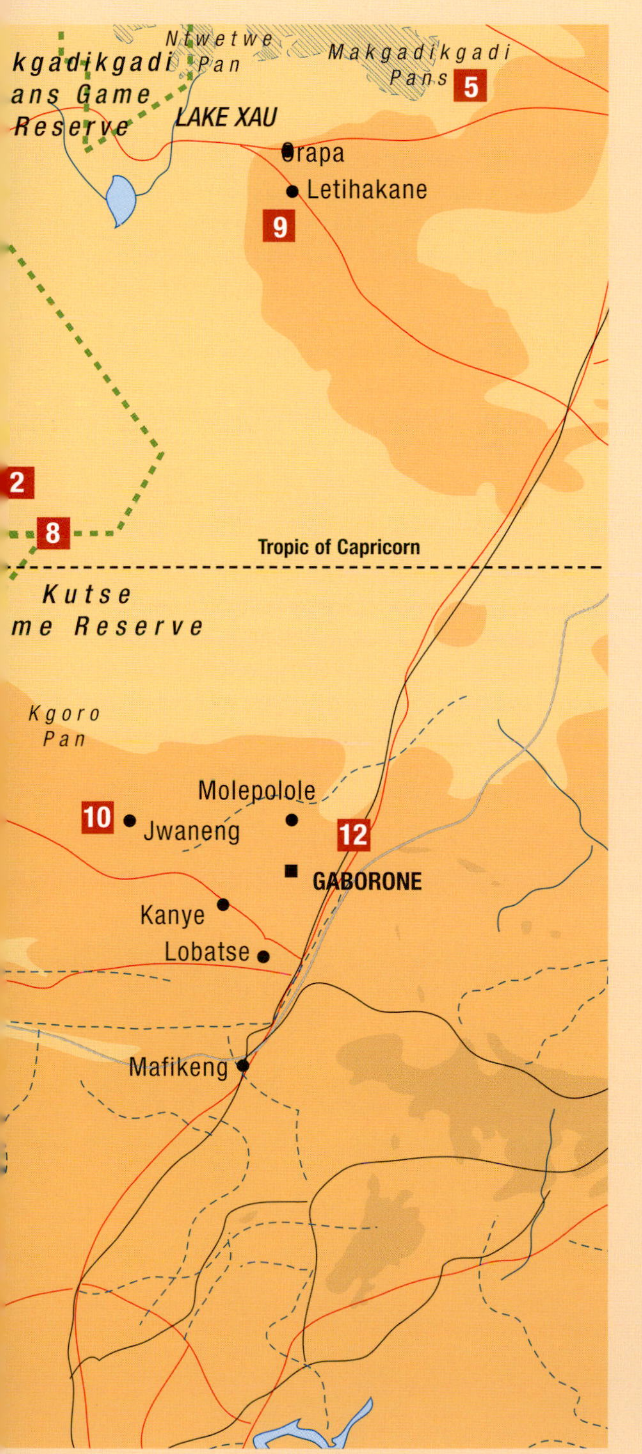


### 5. Makgadikgadi Pans

These pans in northeast Botswana form the largest salt basin in the world (see *illus.*, p. 48). When dry, which is most of the time, it is 6,500 square kilometers (2,500 sq. mi.) of brilliant white salt. When the waters of the Okavango spill down the Boteti River after good summer rains, the whole area of the pan is flooded to a depth of a few centimeters, providing rich feeding for aquatic birds, such as flamingos and pelicans. In recent years Botswana has mined the Makgadikgadi Pans for salt, soda ash, sodium sulphate, and bromides.

### 6. Gemsbok

This remote national park in the southwest corner of Botswana is adjacent to South Africa's Kalahari-Gemsbok National Park. Today the parks form Africa's first transnational park – the Kgalagadi Transfrontier Park. The Gemsbok National Park covers more than 11,000 square kilometers (4,250 sq. mi.) of desert, where the annual rainfall seldom rises above 125 millimeters (5 in.). It is home to huge herds of gemsbok and other antelope, which are capable of living for long periods without water, obtaining moisture from roots and succulent plants.

### 7. CKGR

The Central Kalahari Game Reserve (CKGR) was created in 1961 as a reservation for the San peoples, nomads whose way of life is under threat (*see* pp. 138–139). It is estimated that about 42,000 San live in the CKGR. With an area of 52,000 square kilometers (20,000 sq. mi.), the CKGR is a little bigger than the state of Alabama. There is no standing water in the park, and there are very few tracks.

### 8. Kutse

This relatively small game reserve covers an area of 2,600 square kilometers (1,000 sq. mi.) adjoining the southern border of the Central Kalahari Game Reserve and comprises deep sand and savanna. Lions, leopards, wild dogs, cheetahs, brown and spotted hyenas, and eland are among the wealth of animals to be seen in the reserve.

### 9. Letihakane

This is Botswana's main diamond mine. Begun in 1971, the excavations here and at Jwaneng have helped transform Botswana from one of Africa's poorest nations into one of the richest. The country is the third-largest producer of diamonds in the world.

### 10. Jwaneng

This town sprang up in the 1970s after the discovery of pipes of diamond-rich kimberlite rock 50 meters (165 ft.) below the surface. Now intensively mined by the South African company De Beers, the area is disfigured by slag heaps of discarded materials.

### 11. Woodland

Just over 100 kilometers (62 mi.) west of Tshane are some of southern Africa's most spectacular woodlands, with sumptuous growths of camelthorn trees standing in stretches about 20 kilometers (75 mi.) long and 5 kilometers (3 mi.) wide. Unusually, there appear to be no shrubs or bushes, only mature trees and the grass beneath them.

### 12. Gaborone

The city became the capital of Botswana in 1965, a year before the country achieved independence from Britain. Standing at the edge of the Kalahari Desert on the main railroad between Cape Town, South Africa, and Harare, Zimbabwe, it today boasts a university, an international airport, and several important museums. In 1991 the urban population was about 133,000.

### 13. Ghansiland

The semiarid plateau in northwest Botswana is widely regarded as one of the best cattle-ranching regions in the world. Drought, however, is a constant threat. The regional capital, Ghansi, has one of Botswana's few airports.

## Fact File

- The Kalahari covers an area of about 930,000 square kilometers (360,000 sq. mi.). At its greatest extent it stretches about 1,600 kilometers (990 mi.) from north to south and about 970 kilometers (600 mi.) from east to west. Its boundaries are not always distinct, however, particularly in the southwest, where it merges with the neighboring Namib.

- The Kalahari covers eastern Namibia, part of the Cape Province of South Africa, and almost the whole of Botswana.

- On classification methods based on rainfall, the Kalahari does not count as a desert, receiving almost double the annual maximum of 25 centimeters (10 in.). Other forms of classification, however, allow for the potential evaporation rate of water. If water can evaporate more than twice as quickly as it falls, then a region is classed as a desert. Even so, only the southwestern part of the Kalahari qualifies as desert.

- The Kalahari lies on the interior plateau of southern Africa. The plain, which is a vast basin, is largely featureless, undulating, and covered in sand that is reddish in color. The entire plain lies at least 900 meters (3,000 ft.) above sea level.

- The Kalahari has extreme changes in temperature. Summer daytime temperatures can reach 46ºC (115ºF) but fall to 27ºC (70ºF) at night; on winter nights temperatures often drop below freezing.

# "The Great Thirst"

In strict terms, much of the Kalahari does not qualify as true desert. Although in the southwest the desert receives less than 12.5 centimeters (5 in.) of rain annually, the northeastern region lies in the path of moisture-bearing air from the Indian Ocean and receives twice as much rain as the 25-centimeter (10-in.) maximum often used to define a desert. The rapid evaporation or disappearance of the water, however, creates an aridity as great as that of most true deserts. Almost all of the area's rain falls as summer thunderstorms, but they are highly erratic. In the dry winters no rain falls for six to eight months. It is not for nothing then that the Kalahari's name derives from the Tswana word *kgalgadi*, meaning "the great thirst."

### Geography of the Region

The Kalahari is a vast basinlike plain that occupies around 930,000 square kilometers (359,000 sq. mi.) of the southern African plateau. It is largely flat and featureless, apart from scattered kopjes – small hills with vertical sides that rise abruptly from the surrounding plain. Around 90 percent of the area is covered in sand, which is

**The scrubby landscapes of the Kalahari bear little resemblance to conventional notions of a desert. Despite its relatively plentiful rainfall, however, there is little standing water.**

generally reddish in color. To the north the desert is bordered by the Okavango Delta, or Swamps, and to the northeast and southeast by the Highveld of South Africa. To the south its edge is marked by the Orange (Gariep) River; to the west it runs into the mountains and highlands of Namibia and merges gradually into the coastal Namib Desert (*see* pp. 32–33).

The north is much wetter than the south. The Okavango Delta is fed by runoff water from the Angolan highlands to the northwest. South-flowing streams merge and flow into the northern end of the desert, where they break again into smaller channels and eventually run into a vast wetland area. In the southern and central parts of the desert, on the other hand, virtually all water that falls disappears at once into the sand and there are only a few scattered waterholes. Here a few seasonal pans (locally called *vleis*) are vital for the survival of wildlife as well as the few human settlements. Round or oval in shape, they are usually areas of hard, gray clay, filling the base of shallow natural depression. The western part of the desert is occupied by barchan – or crescent-shaped – dunes.

### Flora and Fauna

Despite its unpredictability, the Kalahari's rainfall means the region is well vegetated. Vegetation varies widely. In the far southwest, low rainfall means that there are few trees or bushes, though some hardy shrubs and grasses do survive. Trees whose roots can tap into permanently moist sublayers of sand can also survive here. Moving northeast the vegetation increases in size and variety, through an area of scattered acacia trees to a region that has little resemblance to traditional notions of a desert. The northeastern Kalahari has open woodlands, evergreen and deciduous forests, and palm trees and is the basis of a considerable timber industry. The neighboring Okavango Delta supports its own wetland ecological system, with reeds, papyrus, and various species of water lily.

Animal life in the region is also concentrated toward the northeast, where the fauna is more reminiscent of that of other African savanna regions, and includes giraffes, zebras, elephants, and buffalo. Big cats and hunting dogs live there, as do a wide range of smaller mammals and reptiles. The trees and bushes are home to many types of birds. In the south, despite the lack of surface water, vast herds of oryx, springbok, and gnu sometimes wander among the scrubby sands. Life survives even in the saline pans in the north, where tiny brine shrimps are specially adapted to the salty conditions.

**Traditionally the San of the Kalahari are nomadic hunter-gatherers. Here a woman digs for roots and tubers, which contain valuable moisture as well as nourishment.**

### People of the Kalahari

The Kalahari is only sparsely populated. Despite its large size (569,582 sq. km.; 219,916 sq. mi.), Botswana has a population of just 1,406,000, most of whom live in the far east of the country. The Kalahari itself supports a small population of Bantu-speaking peoples, including the Tswana and the Herero, as well the San (*see* pp. 138–139). Formerly nomadic, the San today generally live in villages, many of which are in the Central Kalahari Game Reserve.

The mainstay of Botswana's economy is cattle farming, and goats also roam for food around the edges of settlements. In more remote villages other food is provided by wild plants and game animals. Since the opening of a diamond mine at Orapa in the early 1970s, the area has also seen an expansion in mining activity generally, providing a much-needed boost to Botswana's economy.

# Arabian Deserts

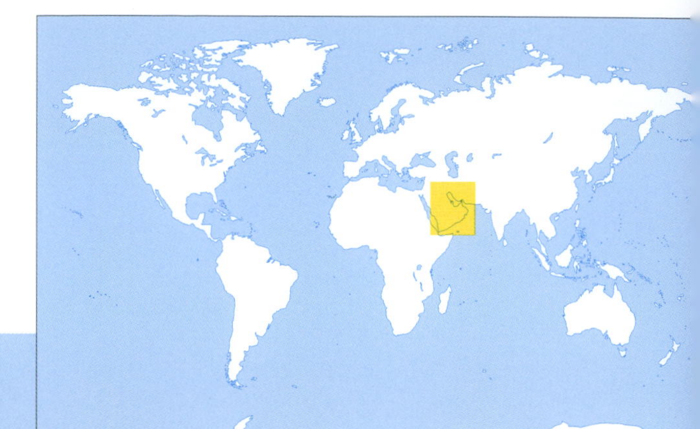

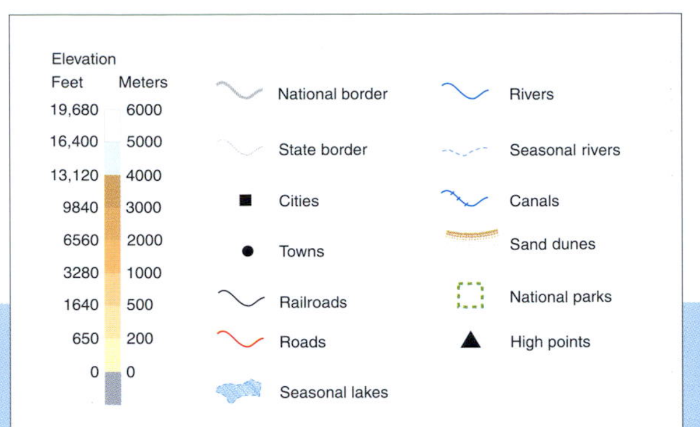


**1. Syrian Desert**
This desert area, including parts of north Saudi Arabia, southeast Syria, west Iraq, and northeast Jordan, is crossed by important pipelines, connecting the oil fields of Iraq with the Mediterranean ports of Haifa and Tripoli.

**2. Palmyra**
This ruined oasis city (*see illus.*, p. 130) on the northern edge of the Syrian Desert (just off map), close to modern Tadmur, was said to have been built by the biblical King Solomon.

**3. An Nafūd**
This desert of red sands occupies a great oval depression in north Saudi Arabia, 290 kilometers (180 mi.) long and 225 kilometers (140 mi.) wide, and is surrounded by eroded sandstone outcrops. The desert is plagued by violent winds, which have formed barchan dunes up to 183 meters (600 ft.) tall. In some lowland basins there are oases, where dates, fruits, and vegetables are grown.

**4. Ha'il**
This oasis town on the northern edge of the An Nafūd was a traditional stopover for pilgrims on their way to Mecca.

**5. Buraydah**
This old trading town in central Saudi Arabia is today home to the world's largest camel market.

**6. Ad Dahnā**
This great arc of sand, about 80 kilometers (50 mi.) wide and 800 kilometers (500 mi.) long, connects the Arabian Peninsula's two great deserts, An Nafūd and Rub' al-Khali.

**7. Oil field**
The Khurays oil field in Ad Dahnā' is one of the largest and most productive deposits of its type in the Arabian Peninsula. It was discovered in 1957.

**8. Riyadh**
This desert capital of Saudi Arabia has a population of 1.3 million (*see illus.*, p. 144). Largely built with money from the oil boom, Riyadh is today a high-tech oasis of glass, steel, and concrete, with one of the biggest airports in the world.

**9. Sebkhas**
These salt pans are a feature of coastal regions of eastern Arabia, where high levels of evaporation of sea water create salt-encrusted plains. The largest exposed coastal formation is Sebkha Matti, which extends for about 97 kilometers (60 mi.) inland.

**10. Medina**
This inland city in Hejaz is the second most holy city of Islam. Muhammad fled here on September 20, 622 A.D. – the date that marks the beginning of the Muslim calender.

**11. Mecca**
The birthplace of Muhammad and the holiest place of Islam.

**12. As-Sab'atayn**
This area of transverse and seif dunes in Yemen covers an area of roughly 25,900 square kilometers (10,000 sq. mi.).

**13. Ma'rib**
This ruined town was a capital of the ancient kingdom of Sheba, famous for it wealth. Ma'rib was the site of a great earthwork dam, constructed in the seventh century B.C. and destroyed in the sixth century A.D.

**14. An-Nabī Shu'ayb**
At 3,760 meters (12,336 ft.), this is the highest peak in the Arabian Desert.

**15. Wadi Hajr**
Apart from the great Tigris and Euphrates river systems at the northern limit of Arabia, this is the only year-round freshwater stream in the whole Arabian Peninsula. It flows for about 97 kilometers (60 mi.) from its source to the Gulf of Aden.

# Fact File

- The Arabian Desert covers an area of around 2,330,000 square kilometers (900,000 sq. mi.) and occupies virtually the entire Arabian Peninsula between the Red Sea and the Persian Gulf. At its greatest extent it stretches 2,250 kilometers (1,400 mi.) from north to south and around 2,100 kilometers (1,300 mi.) from east to west.

- The Arabian Desert is a high-pressure climate desert, maintained by the belt of hot, dry air about and north of the tropic of Cancer (*see* pp. 16–18).

- The desert occupies a large part of Saudi Arabia and parts of Yemen, Oman, the United Arab Emirates, Qatar, Kuwait, Iraq, and Jordan.

- The Arabian Desert is bordered to the east by the Persian Gulf and Gulf of Oman, to the southeast by the Arabian Sea, and to the west by the Red Sea. In the north it merges into the Syrian Desert.

- The peninsula has two major geologic regions. In the west rises the Arabian platform, or Afro-Arabian Shield, of igneous and metamorphic rocks; to the east lie gently sloping flatlands and basins of younger, sedimentary rocks.

- Around a third of the desert is covered in sand, the two greatest areas of which occur in the An Nafūd in the northwest and in the Rub' al-Khali, or "Empty Quarter," in the southeast (*see map*, pp. 40–41). Rub' al-Khali is larger than France.

- An Nafūd has a unique dune form, found nowhere else on earth – a giant crescent slipface (the slope on the leeside of the dune) with a bedrock lee hollow that is devoid of sand. From above the dunes look like giant hoof marks in the desert.

- The nomadic bedouin usually refer to the Rub' al-Khali as Ar-Ramlah, meaning simply "the Sand."

- The first recorded crossing of the Rub' al-Khali by a European was made in 1930–1931 by the British explorer Bertram Thomas, an adviser to the sultan of Masqat.

- Average annual rainfall in the desert is less than 10 centimeters (4 in.) a year, though individual years might see no rain at all or up to 50 centimeters (20 in.).

- The Arabian Desert has some of the richest oil deposits in the world. Saudi Arabia is one of the leading oil producers, in 2002 producing over eight million barrels of oil every day.

## The Desert Peninsula

Almost all of the Arabian Peninsula is covered with desert, an area of about 2,330,000 square kilometers (900,000 sq. mi.). The desert is dominated by two great regions of sand, An Nafūd, or the Great Nafūd, in the north and the awesome Rub' al-Khali ("Empty Quarter") in the south, connected by a long arc of sand, the Ad Dahnā'. Because of the strong seasonal winds that blow across the peninsula, most of the sand forms dunes, of which there are a great diversity. Elsewhere are stony, cherty, or graveled plains – the result of deposits left thousands of years ago by ancient river systems, the remains of which are the wadis that today crisscross the peninsula.

Although large parts of the desert appear entirely barren, it is actually relatively rich in plant and animal life. After rain the central desert becomes home to flowering plants, such as mustard, pea, and iris, while the stony plains and steppes produce grass that provides food for livestock, including horses and camels. Shrubs are often used by the nomadic bedouin to make seasoning, perfume, or herbal cures. Highlands and wadis support juniper and milkweed trees, while date-palms grown in oases are used for building as well as for food.

The most notable of the desert's many animals include many species of scorpion; a toothless lizard called a dab; and highly venomous sand cobras. Resident bird life includes species of falcons and eagles, numerous ravens, and burrowing owls. The mammals, however – ibex, foxes, hyenas, and jackals – have suffered badly from the effects of human hunting, and their numbers have fallen.

In ancient times the Minaean and Sabaean kingdoms flourished in the south of the peninsula, but little is known about them. In the seventh century A.D. Arabia became the birthplace of one of the world's great religions, Islam. The ancestors of the nomadic bedouin (*see* pp. 136–137), who have roamed the desert for thousands of years, were early converts. Many of their modern descendants have now settled in towns or near oases, alongside their fellow Arabs, attracted by the wealth that followed from the discovery, from the 1930s, of the desert's rich oil reserves.

# The Empty Quarter

**1. Climate**
The climate in the heart of the Rub' al-Khali is harsh in the extreme. Daytime summer temperatures commonly reach 50°C (138°F), while at night they often drop to freezing point.

**2. Population**
The Rub' al-Khali is one of earth's least populated areas. Because of this some of the borders between Saudi Arabia and Yemen and Saudi Arabia and the United Arab Emirates remain undefined.

**3. Gravel plains**
The western edge of the Rub' al-Khali is marked by vast gravel plains.

**4. Vegetation**
The Empty Quarter acquires its name not simply because of its lack of human population – only 37 species of vegetation have been identified throughout the area.

**5. Al-Jaladah**
An extensive gravel plain that lies at the heart of the Rub' al-Khali.

**6. Al-Jāfūrah**
This area appears to be a northern extension of the Rub' al-Khali, but is regarded by Arabs as a separate desert.

**7. Dunes**
The largest dunes in the Rub' al-Khali are along its northeastern fringe. They reach heights of 250 meters (820 ft.) and typically extend for 48 kilometers (30 mi.). The longest stretch for up to 240 kilometers (150 mi.).

**8. Sebkhas**
Most of the desert's sand dunes are formed on top of concealed salt flats, or sebkhas. The largest exposed formation of this type is at Umm as-Samīm in the eastern part of the Rub' al Khali. Sebkhas consist principally of a mixture of dune sand and carbonate minerals that have been blown in by the wind from the coast. They are often covered with a surface layer of rock salt.

**9. Oil and gas**
In places the undulating sandscape of the Rub' al-Khali is now broken by drilling rigs, which have been built since 1945 to extract the vast oil and natural-gas reserves.

**10. Minerals**
Parts of the Rub' al Khali are known to contain cadmium, copper, gold, iron, lead, magnesite, platinum, pyrite, silver, titanium, and zinc. Feasibility studies are being carried out to determine if they can be extracted economically.

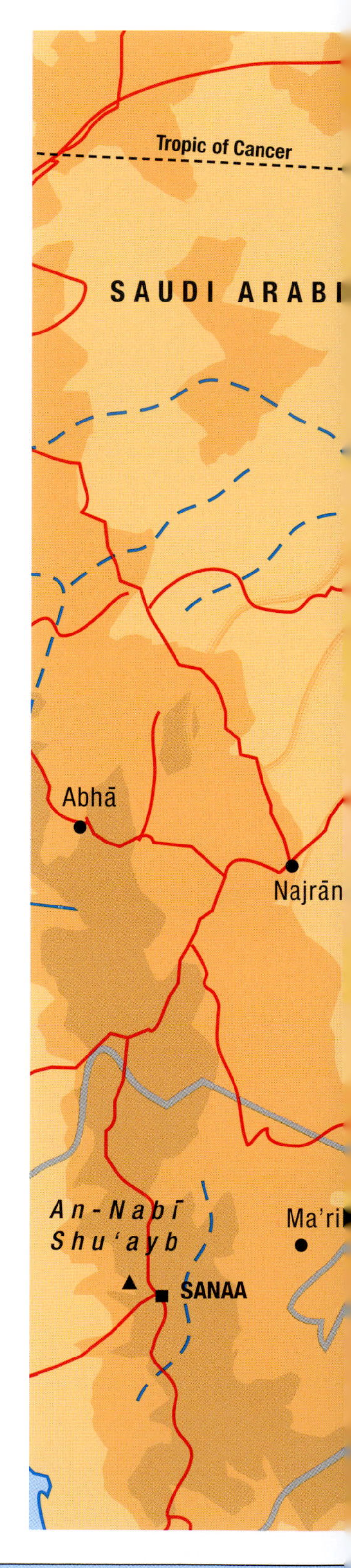

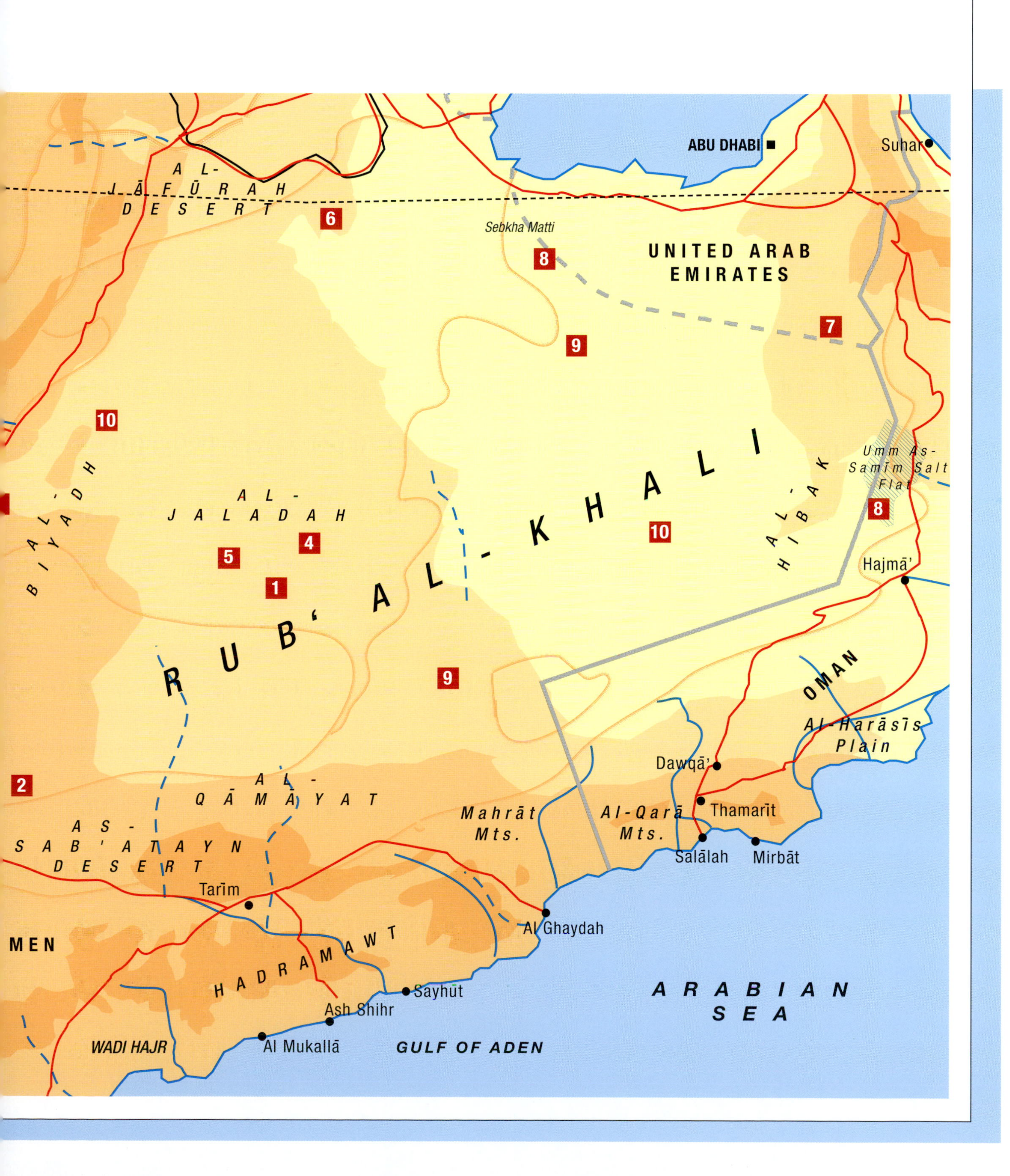
ABU DHABI
Suhar
AL-JĀFŪRAH DESERT
6
Sebkha Matti
8
UNITED ARAB EMIRATES
7
9
10
AL-BIYADH
Umm As-Samīm Salt Flat
AL-JALADAH
5
4
1
RUB' AL-KHALI
10
AL-HIBAK
8
Hajmā'
9
OMAN
Al-Harāsīs Plain
2
AL-QĀMĀYAT
Dawqā'
AS-SAB'ATAYN DESERT
Mahrāt Mts.
Al-Qarā Mts.
Thamarīt
Salālah
Mirbāt
Tarīm
MEN
HADRAMAWT
Al Ghaydah
ARABIAN SEA
Sayhūt
Ash Shihr
WADI HAJR
Al Mukallā
GULF OF ADEN

# SAND, ROCK, AND RUBBLE

*The naked desert terrain is vulnerable to the powerful forces of sun, wind, and water, which over time can slowly create spectacular topographical features such as towering dunes and glistening salt pans. At other times, however, their effects are sudden and destructive.*

The desert landscape is often assumed to be uniform – a barren, featureless wasteland of endless sand or rubble and blistering, blinding sun. Anything more than a cursory glance, however, reveals a much more highly nuanced topography, whose very diversity is made all the more visible by its near nakedness, veiled as it is by only the thinnest cloak of vegetation.

Deserts are among the most eroded landscapes on earth, uniquely vulnerable to the action and interaction of three powerful natural agents: the heating effects of the sun, the scouring power of sand-laden wind, and, perhaps more surprisingly, the occasional onslaught of rushing water. Together, these natural forces carry out the process of mechanical weathering by which rocks are physically riven into ever smaller particles. Wind and water also act as agents of transportation, carrying the particles and debris that result from weathering to new sites. The topography – the surfaces, shapes, and relief of the land – that these natural forces create depends not only on climate but also – as any human sculptor is aware – on the nature of the raw material upon which they work.

In moist, temperate climates two other types of weathering are also important: organic – the breakdown of rocks by animals and plants – and chemical – the decomposition of rocks through processes such as carbonation and oxidation, which remove the "glue" that holds grains together. In arid places, where plant and animal life is sparse and the dry air impedes chemical change, the effects of both these agents is minimal. This can clearly be seen in the remarkable survival of the great granite obelisks and other monuments of ancient Egypt – structures sustained largely thanks to the dry desert air of the nearby Sahara. Once such monuments are moved to temperate-zone cities, however – as in the case of the Egyptian obelisks erected in public spaces in London, Paris, and New York – they quickly register the effects of humidity and chemical decomposition, not to mention the scars of urban pollution.

**An ephemeral river, formed after heavy rains, dries among giant sand dunes in the Namib Desert. As the river evaporates it leaves a whitish salty residue, as visible at the top of the image, while water remains only in the deeper pools.**

## TYPES OF GROUND COVER

Three major types of ground cover occur in true desert regions. The classic sandy surface of hills or dunes, sometimes known as erg, actually overlies only some 20 to 30 percent of desert areas around the world. Some deserts, such as the Great Sandy in Australia and the Namib in southern Africa, are characterized almost entirely by erg. In contrast, sand covers less than 5 percent of the deserts of southwest North America. Other desert areas are made up of gravelly plains strewn with loose stones, pebbles, and other rocky debris, a cover known as reg or desert pavement. This type of ground cover is common both in the Sahara, of which it forms about one-third, and in the Arabian Desert (*below*).

The third type of ground surface, found especially in upland plateaus, is flat, almost bare rock. This is termed hammada and is especially prevalent in Asian deserts such as the Kara-Kum. In addition, some deserts have a thin sprinkling of soil, rarely more than a few centimeters deep and typically pale, coarse, and low in the organic matter known as humus. In most soils humus absorbs and holds water like a sponge, so that low humus content also means low water retention. Because desert rainfall is also insufficient to wash away those minerals that can be detrimental to plant growth, such as calcium and magnesium, many desert soils are both chemically and physically unsuited to plant life (*see also* Chapter 3).

An important general feature of desert erosion is its intermittent nature, contrasting with the generally continuous process found in moister, more equable lands. Months or even years may pass in the desert without hardly anything changing – apart, perhaps, from the occasional crack or shard of rock created under the intense heat of the sun. Then, within days or even hours, the landscape is wrought by violent rains, flash floods, or sandstorms and is transformed, even beyond recognition.

### Effects of Heat and Cold

Because they are often close to the equator and lack cloud cover, many deserts suffer extremely high temperatures as the sun blazes down day after day. However, the absence of cloud cover also means that heat disperses rapidly at night, leaving the desert surface thoroughly chilled. The resulting extremes in temperature rival those anywhere on earth. In the Gobi Desert, for example, summer air temperatures may exceed 40°C (104°F) at midday but fall below 10°C (50°F) at night. Winter air temperature ranges are even more dramatic. Winter nights in the Gobi may plummet to a bone-chilling -30°C (-22°F), making this desert the coldest outside the polar regions.

Ground temperatures are generally much higher than air temperatures. Thermometer readings from the Kara-Kum show that when the midday air temperature is 45°C (113°F), the ground can heat to a searing 80°C (176°F). Only some 12 hours later the ground temperature readings have plunged to 10°C (50°F). Such drastic falls and rises of temperature every 24 hours play havoc with solid rock, its surface alternately baked and almost frozen. Because most rocks are made up of a mixture of minerals, each with its own coefficient of thermal expansion – the amount it expands for a certain temperature rise – the diurnal alternate heating and cooling of its component minerals gradually weakens the rock, causing microscopic cracking. Over time this causes thin sheets of rock to flake, peel, and split off – a process known as onion-skin weathering. Water from rare dews, frosts, or rainfalls can speed up this weathering process if temperatures drop below freezing: the water soaks into tiny pores in the bedrock and cracks and then expands as it turns into ice, prising open the existing crevices.

The Gobi's surface of pebbles, coarse gravel, and salty sand is a typical result of such thermally induced weathering. The region's strong winds blow away fine particles almost as they form, leaving behind in the main the coarser particles, which are smoothed by the departing windblown debris.

## Effects of Water

Deserts are not entirely waterless, and the briefest perusal of any of the maps in this book will reveal any number of rivers, lakes, and oases that appear in what is otherwise dry land. While many of these bodies of water are seasonal or ephemeral, the distribution and circulation of water – the desert's hydrology – play as crucial a role in the shaping of the land as they do in other biomes.

The typical desert contains two types of water. Surface water is extremely limited in quantity and area, may be temporary or permanent, and includes seasonal lakes and streams and the rare mighty river as well as violent flash floods. Groundwater, as its name suggests, is found underground, making itself known in the form of sporadic springs and oases or as artificial wells and boreholes.

Permanent, or at least semipermanent, rivers arise outside many deserts, often in an upland catchment area, and flow across the dry land on their way to the sea or large inland lakes. Such rivers are called exotic streams, and examples include the Nile and Niger in the Sahara, the Tigris–Euphrates in the Syrian–Arabian region, the Tarim of the Taklimakan, the Kerulen of the Gobi, and the Colorado of the Sonoran Desert. Only a large flow of water can cross the desert without drying out.

Occasionally waterways enter a basinlike desert and, finding no natural outlet, spread out to form shallow lakes from which the water either slowly evaporates in the heat or trickles into the ground. As more water continues to arrive, loaded with salty minerals gathered from rocks earlier in its travels, the minerals in the basin gradually become increasingly concentrated. The almost inevitable result is salty lagoons or, in the dry season, glistening salt flats or pans.

### Ephemeral rivers

Many deserts are crisscrossed by the channels of ephemeral rivers, carrying water only occasionally and fleetingly. Sometimes the channels are permanently dry and mark the remnant riverbeds of long-extinct waterways. Usually steep-sided and flat-bottomed, the channels of both permanently dry and ephemeral rivers are called wadis in Africa and East Asia and arroyos in parts of the Americas.

## "PLACE OF NO RETURN"

The western region of the great Kara-Kum Desert in Turkmenistan, next to the Caspian Sea, is overlooked by the high, mainly limestone hills of the Ustyurt Plateau. It is one of the most desolate places in the world, known locally as Barsa Kel'mes, or the "Place of No Return." In this alien landscape of mud, clay, sand, silt, and salt, plagued by tectonic instability, incessant winds, and rare flash floods, almost no life survives.

The region's debris of stones and pebbles mixes with the sand, mud, and tiny salt crystals that blow across from the Caspian Sea. Groundwater laden with dissolved salty minerals is also drawn from far below up to the surface by capillary action. As the water evaporates, it leaves behind salt on the surface, which during rare rainfall mixes with the desert sands and clays to form a paste that quickly hardens in the blazing sun.

Some of the region's most spectacular features, however, are formed by volcanic action. Barsa Kel'mes lies at the edge of one of the planet's tectonic plates and is crossed by major cracks and faults. Its "mud volcanoes" are craters containing mixtures of oil and sand that bubble up from deep within the earth. Scattered between the craters are pools of red water, also drawn from far below the desert surface, which erupt like thin-lava volcanoes. The water evaporates quickly or trickles back down among cracks in the rocks, leaving its iron-oxide minerals to stain the desert surface rust red.

Bowl-like lumps of coarse gravel are often visible in their banks, deposited during brief periods of intense flow. Many of these channels are small and retain water in pools into the dry season and damp soils for even longer, providing shady, moist havens for wildlife and human travelers alike. However, any plants that do take root in these transient waterways must be able to resist the destructive power of the flash floods that sweep through periodically.

Sometimes permanently or semipermanently dry river channels are major landscape features. The Wadi 'Arabah, for example, forms the natural eastern boundary of the Negev Desert in Israel, stretching 150 kilometers (94 mi.) between the Dead Sea and the Gulf of Aqaba. The associated trough continues through the Red Sea into East Africa as part of the Great Rift Valley system. In the Kara-Kum the great, stony northern flatland of the Zaunguzk Plateau descends a steep escarpment to the erg farther south. The scarp marks the one-time course of the great Amu Dar'ya River, which massive movements of the earth's crust have shifted far to the north and east. In the central Negev Desert, at altitudes of about 1,000 meters (3,280 ft.), curious long and thin crater formations have been eroded into the domed, upfolded layers of rock by long-disappeared rivers. They are called *makhteshim*, the largest of which can measure more than 30 kilometers (19 mi.) long, 8 kilometers (5 mi.) wide, and more than 300 meters (984 ft.) deep.

## Water shapes the land

Although water is one of the major landscape-fashioning agents, by itself it does not possess great erosional power. Its destructive force comes, rather, from its load – the particles of rock that its current lifts and carries along. The faster the flow, the larger the pieces transported – from tiny pieces of mud or silt in slow-moving currents up to massive boulders in the fastest floods. The particles bounce and crash against the bottom and sides of the channel, cracking and chipping off more bits and pieces. When the current slows, the particles are deposited in reverse order, the biggest first. In this way water not only transports, but also sorts. As it slows, a fast stream drops first its pebbles, then gravels, sands, and clays in graded fashion as its momentum fades.

This effect can be clearly seen in some deserts bordered by mountains, such as the Patagonian Desert, whose western reaches merge into the foothills of the Andes. On the higher slopes rushing mountain rainwater carries eroded debris downward into short, deep, steep troughs and canyons. As the gradient lessens and the trough widens, the coarse

**In Arizona's Petrified Forest National Park, arroyos, or water-carved channels, cut through the desert floor far below a high plateau. After rare heavy rains water rushes along the arroyos, depositing sediments of sand and gravel on the desert lowlands.**

# THE RAINBOW BRIDGE

One of the most spectacular water-eroded features of the desert is the impressive rock arch in southern Utah known as the Rainbow Bridge (*below*). This natural wonder is situated on a remote desert plateau, embedded among the stream-carved arroyos that wind down from the northern slopes of nearby Navajo Mountain, and can be reached only on foot, by horseback, or by boat on Lake Powell. Originally undercut from a spur of sandstone by a river that once flowed over the site, the bridge is probably the world's largest natural bridge, spanning 84 meters (276 ft.) and towering 88 meters (289 ft.) high. At the top the arch is 13 meters (43 ft.) thick. The bridge's predominant colour is salmon-pink, with dark stains called desert varnish that are caused by iron oxide (hematite). The feature is home to numerous plants, including wild orchids and maidenhair, which are fed by water seeping from the arroyo walls.

The bridge has inspired awe and wonder throughout time. For local Navajo people it is a sacred place, said to have been created when a spirit threw a rainbow across a flood in order to rescue another spirit.

debris is deposited first, then smaller particles, in a spreading conelike shape termed an alluvial fan that extends out into the desert plain. A long mountain range may have many fans at intervals that merge to form a lengthier, broader feature called an alluvial apron, or bajada.

A related erosional feature is the pediment, which is also common around lower slopes of mountains in or fringing deserts. This results from water runoff that is unchanneled by canyons and valleys, called sheet wash, which creates a flat plain or similar broad surface of low relief. The pediment may be bare bedrock or it may be covered with a thin layer of sand, gravel, or similar alluvium. Pediments develop especially between mountains and basin areas.

The landforms described above are easily identified in arid regions, partly because of the lack of vegetation cover. They are usually created quickly by rare rains, producing not the smooth, sloping shapes found in damper climates, where there is more constant wear to the rocks, but angular outcrops and irregular terrain. There they remain for years or centuries as graphic illustrations – almost real-life diagrams – of the erosional process.

**An alluvial fan, marked by the darker color of its sediment, spreads into Death Valley in California.**

An aerial view shows the glistening Makgadikgadi Pans in northeast Botswana, the largest salt basin in the world. In rainy seasons the pans flood extensively and become home to large numbers of flamingoes.

## Lakes wet and dry

There are all kinds and shades of lakes in deserts. Many are temporary, formed by water that falls as rain over nearby uplands and subsequently drains down gently inclined slopes, or bahadas, and into a low point or depression or basin, in the usual way that lakes form elsewhere. As seasonal lakes dry out in the heat, their dissolved minerals crystallize, forming hard crusts of clay, salt, or sometimes both. These are salt lakes and pans, sometimes known as playas or salars. Their genesis is in repeated flooding and evaporation that have taken place over thousands of years.

A true playa – the term derives from the Spanish word for "shore" or "beach" – is an arid-zone lake bed that has no outlet. Occupying the lowest part of the local topography in an enclosed basin from which water cannot flow away, its moisture is lost entirely by evaporation. Heavy storms flood the area to create a shallow lake, but the water lasts just a few weeks. The concentration of salts when the area is watery, and the hardness of the often clay-rich soil when it is dry and baked, means that plants have a very tough time. After each flood the surface is renewed, hard, and perfectly flat – and ideal for human pursuits such as land yachting or vehicle speed trials. Many playas also contain valuable mineral deposits (*see* p. 172).

In southern Africa the Makgadikgadi Pans of northeast Botswana cover a huge part of the Kalahari Desert. They are remnants of the ancient Lake Makgadikgadi, which three million years ago, when the climate was damper, was probably one of the largest expanses of freshwater in Africa. Today the crusty, blinding white plains are the world's largest example of salt pans, covering almost 40,000 square kilometers (1,550 sq. mi.).

## Flash floods

One of the desert's most cataclysmic events is the flash flood, or cloudburst flood, its awesome power able to wash away not only sand, soil, and surface debris, but also deep-rooted plants, boulders, buildings, vehicles, and bridges. Even locomotives on transdesert railroads have been picked up like toys by the surging flood's massive kinetic energy and transported several kilometers. The rushing water sweeps the ground clean, carves new channels, and dumps its load at a new site, altering the entire appearance of the land within hours.

Floods occur almost everywhere, but it is in the desert that their devastating effects are swiftest and most drastic, owing principally to the climatic conditions and the nature of the terrain. The aerial culprits are high-intensity convectional nimbo-cumulus clouds – summer thunderstorms – which

In the wake of a cloudburst water collects in troughs between sand dunes in the Namib Desert. Flood waters evaporate rapidly in the hot desert sun, often leaving behind salt pans.

## THE "PORCELAIN DESERT"

The White Sands National Monument, part of the Tularosa Basin in the U.S. state of New Mexico, covers an area of some 700 square kilometers (270 sq. mi.) north of El Paso between the San Andreas Mountains to the west and the Sacramento Mountains further east. Dubbed the "Porcelain Desert," this national monument is renowned for its soft, snow-white sands, of which the grains are not formed from silica minerals (silicon oxides) like ordinary sand but from gypsum, or hydrated calcium sulphate – the basis for plaster of paris. The sands are also cool to the touch because they reflect much of the sun's radiation. The "Porcelain Desert" is the world's largest surface deposit of gypsum.

As in a sandy desert the grains pile into dunes up to 15 meters (50 ft.) tall, which the wind may displace some 5 to 8 meters (16–26 ft.) each year.
In such dry, high-alkaline soil only a few plants, such as yuccas, sumac, and cottonwood, can take root. The desert is also home to the White Sands Missile Range, a testing facility for the U.S. military.

often build up around topographic features such as hilly outcrops or ravines where convectional air currents are strongest. They rarely last more than a few hours, but within this period a remarkable 25 centimeters (10 in.) of rain can fall. The water hammers onto the ground, but because this is often solid rock, or clay-based soil baked hard and impervious by months of dry heat, there can be as much as 90 percent runoff.

The speed, spread, and direction of the flood depend on the local lie of the land. As the nearby channels and gullies are inundated, the water typically picks up local sediments and develops a leading edge that flows as a dense, viscous, almost syrupy wall up to 1.5 meters (5 ft.) high. On a steep slope this can easily outpace a human, moving at more than 30 kilometers per hour (19 mph). Tremendous erosion occurs as the land is scoured bare. Plants are uprooted and animals drowned in their burrows or on the run, leaving behind a feast

for desert scavengers such as vultures, condors, jackals, and hyenas.

One of the greatest floods in arid regions may have been recorded as the Great Flood in the biblical book of Genesis, as well as elsewhere in Near Eastern legend and literature, such as the story of Gilgamesh. Archeological evidence suggests a severe flood did indeed occur along the Euphrates River some 5,000 years ago, inundating the Ur region of southern Mesopotamia. The devastating effects of arid-zone flooding were witnessed more recently in 1988, when torrential rains flooded the Nile in Sudan, making some 1.5 million people homeless in the area around Khartoum. Such a regional cataclysm makes it easier to understand how an ancient flood might be understood as a worldwide apocalypse.

### Oases: jewels of the desert

The sparse vegetation and aridity of some deserts are occasionally broken by the eruptions of lush greenery and freshwater known as oases. Here, permanent plants such as palms are able to flourish, supporting small-scale human habitation and

## JEWEL OF THE KALAHARI

The Okavango Delta of northwest Botswana is a vast area of lakes, swamps, and reed beds, some 1,610 kilometers (1,000 mi.) wide, in the northern stretches of the Kalahari Desert. It is essentially one giant oasis. The source of the delta's water is the Okavango River, which begins on the Bié Plateau in Angola (where it is known as the Kubango River), flows south and then southeast across the Caprivi Strip in Namibia before emptying into the delta. If this entirely inland drainage system had no outlet, the waters would become saline due to evaporation. There are, however, outflows to the southeast toward the Makgadikgadi Pans and occasionally northeast toward the great Zambesi River and Lake Kariba. The outlets take away on average only 3 percent of the Okavango's inflow, but this is sufficient to keep the water clear, fresh, and sparkling.

During dry periods the Okavango Delta extent forms a triangle of about 15,000 square kilometers (5,800 sq. mi.), but in wetter years this area is liable to swell with the annual flood until it covers more than 22,000 square kilometers (8,500 sq. mi.). The region is a rich haven for wildlife, with the perennial lake areas dominated by phoenix palm and papyrus (*Cyperus papyrus*). Large grazers, such as elephants, zebras, buffalos, wildebeests, and giraffes, mingle with true swamp creatures including crocodiles and many kinds of frogs and fish.

lightest ones farthest. This is why each area of sand dunes has well-sorted grains – they are all very similar in size.

Geologists usually define rock particles by their size in diameter measured in millimeters, from the largest at 2.0 millimeters to the smallest, at 0.1 millimeters. Dunes are composed mainly of grains within the size range of 0.1 to 0.5 millimeters. If rock particles are smaller than about 0.1 millimeters, they tend to stick or clump together, usually forming a coherent, smooth surface that the wind races over without disturbing the uppermost layer. On the other hand, particles larger than 0.5 millimeters across are too heavy for most steady prevailing winds to pick up.

The particles of sand move and sort themselves in three different ways:

• In suspension, as clouds of fine, dustlike particles that can rise to altitudes of more than 2,000 meters (6,560 ft.). In particularly volatile conditions suspended sand particles can result in sandstorms (*see* pp. 56–57).

• By saltation, that is, jumping along the desert surface like a bouncing ball. This is the most common form of motion in dune-building.

• By rolling. Larger grains, too heavy to become really airborne, are nevertheless swept across the desert surface.

## DASHT-E-LŪT

One of the two major regions of the Iranian Desert, Dasht-e-Lūt in central and east-central Iran is a large, dry, hot basin where rivers running off the nearby mountains end their journeys. The region is about 500 kilometers (310 mi.) by 300 kilometers (186 mi.). In the west, at the foot of the Kermān Mountains, it is mainly rocky. In the wet spring rivers bring down pebbles and gravels to form deltalike deposits known as *dasht.* On the eastern side are low, flat salt pans, but toward the center of the basin rock and sand mix with clay and salt after rare rain to form a thick paste that dries to a crust and is scoured by the strong winds that blow almost constantly from the northwest. The results are jagged ridges, some 70 meters (230 ft.) tall and stretching for more than 100 kilometers (62 mi.), with troughs between. Sharp-edged pyramidal points and ravines are scattered among them, creating an impression of utter desolation. To the southeast of Dasht-e-Lūt lies a vast sand field, or erg, where some of world's tallest dunes tower more than 300 meters (984 ft.).

**Wind is a powerful weathering agent, carving out spectacular rock formations, such as these craggy towers in the Sahara, Algeria.**

**The Namib Desert encompasses some of the most striking sand formations in the world, including areas of towering star dunes and barchans, or crescentic dunes. Some of the tallest dunes in the world are found close to the Sossusvlei salt pan.**

Accumulation of sand particles tends to occur where the land dips or shows other slight irregularities in topography, or where features such as boulders or vegetation project above the general surface level. Once started, the dune itself becomes such a topographical feature and continues to build. Studies carried out in the Sahara showed that a dune 3 to 4 meters (10–13 ft.) high and about 100 meters (330 ft.) long needs perhaps 40 to 50 years to form.

Sand dunes are not always stable, but may shift across the landscape. If the prevailing wind is steady, the dunes may move, or "creep" along, in the same direction as the breeze, by as much as 20 to 30 kilometers (12–19 mi.) per year, but usually much less. The tallest "desert waves" may reach as high as 500 meters (1,640 ft.).

## Types of dune

There are several schemes for grouping sand dunes by their overall shape and arrangement, but a number of basic patterns are generally recognized and these are outlined below and in Figure 2 (*opposite*). In reality the numerous variables – particle size and density, wind speeds and directions, and underlying topography – mean that dunes grow, shift, merge, split, and fade in all kinds of mixtures and combinations.

• Linear, or seif (from the Arabic word for "sword"), dunes are strait or slightly sinuous sand ridges that may be up to 300 kilometers (186 mi.) in length. A linear dune may be isolated, but it more often forms part of a set of parallel ridges, divided by interdune corridors of sand, gravel, or rock. Linear dunes tend to form where sand is scarce and the winds are strong and steady, as is the case with the trade winds.

• Crescentic, or barchan, dunes are the most common form of dune. When seen from above they are C-shaped, with the center of the curve perpendicular to the prevailing wind, which is usually regular and steady in speed. They usually form in groups, on smooth hard ground with a

## THE GREAT ERG OF BILMA

Part of the Sahara, the Great Erg of Bilma fans out southwest from the Tibesti Mountains into Chad and Niger. This massive sand field is some 1,200 kilometers (745 mi.) long and averages 150 kilometers (93 mi.) in width. The erg's grains are supplied from the Tibesti by wind and water erosion, and the prevailing wind, the harmattan, blows steadily from the northeast for eight months of the year. The result is an exceptionally rich texture of different dune types.

At the erg's center is a fairly flat sheet, while toward the east barchans predominate. Star dunes, or ghourds, occur in various places, with open areas of barer rock called *gassis* between them. In the south are mainly seif dunes, some more than 150 kilometers (93 mi.) in length and 1,000 meters (3,280 ft.) wide, lying generally parallel to the harmattan. At the heart of the erg is the oasis town of Bilma, once an important stop on the trans-Saharan trading route between Central Africa and Libya, but today difficult to access.

limited sand supply. The ridge may be cusped, with a shallower slope on the windward side where grains roll up and then tumble over and down the leeward side. This may be scalloped or scooped out by the wind's swirls and eddies as it goes over the crest. Barchans can be "seeded" by a boulder or plant that stabilizes a patch of sand and which then becomes the dune's central area. Its tips, or "horns," grow and lengthen, pointing downwind. Unanchored barchans march across the desert at speeds of some 20 to 30 meters (65–100 ft.) per year, with bigger specimens traveling even faster. Sometimes several group into a V-formation, with the point of the V facing the wind – a phenomenon sometimes likened to a flock of low-flying giant geese passing majestically across the landscape.

• Star, or radial, dunes are created by varying, multidirectional winds and are the dominant feature of the Great Eastern Erg, one of the major sand sheets of the Sahara. They also occur in other deserts, usually around the margins of sand seas, near topographic barriers. They have arms of differing lengths radiating from a central mound, and once formed are fairly stationary, although they may become skewed or distorted if the prevailing winds become more regular from one direction for a time. Because they grow upward rather than laterally, they are often very tall. In the Badain Jaran Desert of China, star dunes tower 500 meters (1,640 ft.) into

**Figure 2 Top to bottom are seif, or linear, dunes; crescentic, or barchan, dunes; and star, or radial, dunes. Dunes are usually classified into one of six major types, with many complex or compound forms also occurring.**

**Crescentic, or barchan, dunes cover Niger's Ténéré Desert, part of the Sahara. Barchan dunes form where the prevailing wind blows from one direction – from left to right in this photograph.**

the air and may be the tallest on earth. Some star dunes are rounded, while others have more pyramidal central mounds, with flatter sides and sharper angles. Closely neighboring star dunes can sometimes merge to create a more random pimpled or pockmarked effect.

• Parabolic, or U-shaped, dunes – most commonly found in coastal deserts – tend to build under slightly more humid conditions, especially if vegetation holds back the side arms while the central part of the dune continues to march downwind. The longest known parabolic dune has a trailing arm 12 kilometers (7.5 mi.) long. Over time the U gradually elongates and the middle section may eventually blow out to leave two parallel ridges.

• Sand sheets or stringers are flat regions where sand grains are in plentiful supply and winds are light and variable. They are the desert equivalent of a flat calm on the ocean.

In general simple dunes of each type are formed where the wind regime has not changed in intensity or direction since the formation of the dune. Complex dunes – in which two or more types combine – indicate that the intensity and direction of wind has changed.

### Sandstorms and dust storms

Apart from the flash flood, the other great cataclysm of the desert is the sandstorm. This occurs when grains of sand are caught up and swirled through the air by strong convection currents. Such storms vary hugely in size and duration.

Small tornadolike air movements just a few meters across skip across the desert, whipping up particles in their turbulent spirals, and are visible from a distance of a kilometer or more. These mini-whirlwinds are known by various names such as "dust devils" and, in Australia, "willy-willies." Few, however, last more than 30 minutes. Strong winds, such as the haboob of Sudan, raise larger sandstorms where the average grain size is 0.8 to 1.0 millimeters in diameter. The strong air currents carry the particles in suspension or bounce them along the surface, usually less than 3 meters (10 ft.) off the ground and rarely above 15 meters (50 ft.). They swirl with the winds as convection currents of hot air rise, cool, and then fall. It may take an hour or two for the storm to pass or dissipate.

At the largest end of the range of sandstorms a vast mass of air over an arid area, heated by the sun and by reflected heat from the ground, may suddenly rise to a height of some 10 kilometers (6 mi.) or more, jerking upward like a hot-air balloon released from its mooring ropes. Cooler air moves in rapidly from the sides to replace the rising, staying low as its tonguelike eddies lick along the ground. It, too, is warmed by the hot ground and begins to pick up dust, silt, and fine sand which it swirls along. Such events, sometimes called dust storms, are especially common in the Sahara, where intense heat and strong winds coincide

**A sandstorm blows up in the Kalahari Desert, in southern Africa. Sandstorms are a natural hazard of desert regions, reducing visibility and burying buildings, vehicles, and cattle.**

## THE GIANT'S PLOWED FIELD

An especially impressive display of linear, or seif, dunes is a feature of the Gibson Desert, which lies in Central and east Western Australia. These spectacular sand formations run generally parallel to one another in a northwest-to-southeast direction and are relatively stationary. Most of the dunes are about 1,000 meters (3,280 ft.) wide and 30 meters (98 ft.) high, but they stretch for distances of over 200 kilometers (124 mi.). The oxidized, or "rusty," iron minerals in the sand grains impart distinctive flaming reds and glowing oranges to the landscape. Tough, spiky spinifex plants help to anchor some of the dunes in place, in the same way that marram grass stabilizes dunes along the coast. From high above in aerial photography, they look like the desert version of corrugated cardboard, or as if a giant had plowed a sandy field.

during the summer period of atmospheric instability. These hot Saharan dust storms are known as simooms.

The larger particles of sand are concentrated near the ground but huge quantities of dust can shoot up to 4,000 meters (13,120 ft.) in altitude, posing a serious hazard to passing aircraft. The winds attain average speeds of 50 kilometers per hour (30 mph), occasionally gusting to 100 kilometers per hour (62 mph). The cooler air also causes temperatures to plummet by 12 to 15°C (22–26°F) in a few minutes, and the windblown dust and sand may be so dense that visibility is reduced to zero for as long as three hours. As the storm passes or abates, it leaves sand and silt up to 3 meters (10 ft.) deep, piled in random heaps or against rocks, plants, vehicles, and buildings. Many desert-dwellers tell stories of people, traveling caravans, or even whole villages being smothered and swallowed by such raging sandstorms.

## MIRAGES

Though popularly associated with the desert, mirages occur in both hot and cold climates, and both on land and sea. They belong to a group of natural optical phenomena that includes the fata morgana, the ignis fatuus (will-o'-the-wisp), and the Brocken specter. The most common desert mirage is of a large, distant sheet of water shimmering in the sunlight. It occurs when air close to the ground becomes so hot that it loses density in comparison with the layers of air above it. Rays of light passing through the air are bent, or refracted, and may strike the eyes of an observer as distorted images of the sky or the desert surface that can easily be mistaken for water. In exceptional circumstances other images – lush oases or even cities – may be visible.

Mirages can also be observed in polar deserts. Light rays traveling from a warm layer of air resting on a colder layer, which has been cooled by the underlaying ground surface, cause a double inverted image of a distant object to appear to float over a flat area of snow or ice.

KARA-KUM
GOBI
TAKLIMAKAN
IRANIAN
THAR
ARABIAN
INDIAN OCEAN

# ASIAN DESERTS

*Covering some 44,391,162 square kilometers (17,139,445 sq. mi.), Asia is the world's largest continent, with vast inland areas that are hundreds and sometimes thousands of miles from the sea. It is this "continental effect," together with Asia's soaring mountain ranges, that maintains most of the continent's arid regions, including the awesome Kara-Kum, Taklimakan, and Gobi deserts.*

For the Mongolian nomads of the Gobi Desert, horse riding was – and continues to be – a vital and highly prized skill. These young Mongolian boys are competing in a horse race as part of the celebrations for Mongolia's National Day.

# Kara-Kum Desert

**1. Aral Sea**
The reduction in flow of both the Amu Dar'ya and Syr Dar'ya rivers, following massive irrigation projects begun in the late 20th century, have drastically reduced the size of this inland sea. Once the fourth-largest inland body of water in the world, it has today lost more than three-fifths of its volume of water (*see* p. 183 *and illus.*).

**2. Kyzyl-Kum**
This large desert (about 300,000 sq. km.; 115,000 sq. mi.) lies between the Syr Dar'ya and Amu Dar'ya rivers, southeast of the Aral Sea. Comprising a plain of sand ridges and a few mountain peaks, the desert serves as pastureland for sheep, horses, and camels. Natural-gas deposits are exploited at Gazli in the south of the desert.

**3. Amu Dar'ya**
This long river (2,538 km.; 1,578 mi.), the ancient Oxus, flows from the Pamirs to the Aral Sea, its lower course running largely through desert. Its waters feed a series of oases along its length and support some of Turkmenistan's and Uzbekistan's most important cities. The diversion of its waters into the Kara-Kum Canal for irrigation purposes has drastically reduced its flow while increasing its salinity, and its once large delta at the Aral Sea is all but dried up.

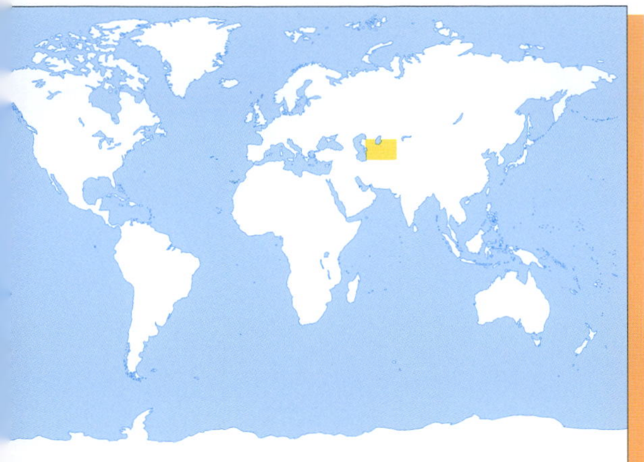

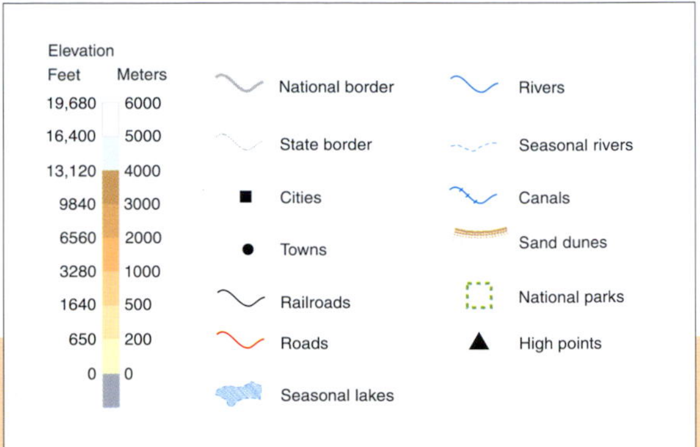

**4. KARA-KUM CANAL**
The Kara-Kum Canal is one of the world's largest irrigation and shipping canals. The canal, which was begun in 1954, stretches from the Amu Dar'ya almost to the Caspian Sea, skirting the southern fringe of the Kara-Kum Desert, and is some 1,400 kilometers (870 mi.) long. Work is in progress to continue the canal to Kizyl-Atrak on the Caspian Sea.

**5. KOPET-DAG OASIS**
This oasis stretches along the northern foothills of the Kopet-Dag Range, on the southern fringe of the Kara-Kum. Both the mountains and their foothills are also rich in mineral resources, including oil. The economic and cultural center of the oasis is the Turkmen capital Ashkhabad.

**6. MURGAB**
The Murgab River flows west and northwest from the Hindu Kush, before petering out in the sands of the Kara-Kum. Its waters feed the Murgab Oasis, the largest fertile area in the desert, famous for its cotton, silk, rugs, and carpets.

**7. MARY**
This ancient town, formerly known as Merv, is the largest settlement in the Murgab Oasis and was traditionally considered to be the site of Paradise. In addition to being an important cotton-growing center, it has a natural-gas power plant and factories producing oil-drilling equipment.

**8. KHAUZ-KHAN**
The reservoir built here after the completion of the Kara-Kum Canal brings much-needed water to the surrounding area.

**9. ASHKHABAD**
The Turkmen capital was almost completely destroyed in 1948 in one of the earthquakes to which this region is prone. Today the city's main industries are glassmaking and carpet weaving; and there are also numerous cotton mills. The city's beautiful setting has made it a favored location for film makers, many of whom now have offices in the city.

**10. SILK ROAD**
Numerous Central Asian towns and cities, such as Mary, Samarqand, Bukhara, and Tashkent, were important trading places on the ancient caravan routes between China and Europe.

**11. PIPELINES**
The desert is crisscrossed by a growing network of pipelines transporting the high-grade oil from drilling rigs to refineries near Ashkhabad. The Kara-Kum's oil and natural-gas deposits are crucial to Turkmenistan's future prosperity.

## Black Sands

The Kara-Kum Desert covers almost 90 percent of the Central Asian republic of Turkmenistan. To the south its fringes run parallel to the Kopet-Dag Mountains; to the north and east lies the great Amu Dar'ya River, which runs down from the Pamir Mountains; while to the west the desert almost reaches the shores of the Caspian Sea. The desert owes its names to its grayish sands – *Gara Gum* means "Black Sand" in the Turkmen language. Across the Amu Dar'ya is another great region of desert, the Kyzl-Kum ("Red Sand") Desert of Uzbekistan and Kazakhstan.

Characteristic of the Kara-Kum, particularly in the low-lying plain of the central region, are half-overgrown sand ridges that may reach up to 90 meters (300 ft.) tall. To the southeast are areas of salt marsh, while to the north is a region of windswept plateau, including Barsa Kel'mes, or the "Place of No Return" (*see* p. 45).

Oases cluster around the Amu Dar'ya and smaller rivers. Large clay-lined hollows called takyrs act as natural reservoirs during times of rain, helping to support the Kara-Kum's surprisingly varied vegetation, including grasses, bushes, and shrubs; some drought-resistant trees even grow among the barchans.

**Most of the sand dunes and ridges of the Kara-Kum are lightly vegetated with grasses and scrubby bushes and shrubs, watered by scattered springs and takyrs.**

By contrast, the desert supports limited wildlife. There are few large mammals, although a species of gazelle occurs, as do barchan cats and corsac foxes. Insects include numerous types of beetles, and as in all deserts reptiles are represented by species of lizard and snake. The bird life includes desert sparrows and saxaul sparrows.

Like other deserts, the Kara-Kum underwent massive changes in the 20th century. Before Turkmenistan became part of the Russian Empire in the 19th century, the Turkmen were largely nomadic pastoralists, raising cattle, camels, sheep, and goats around the desert's oases and takyrs. In Soviet times the region was intensively developed, and the nomadic lifestyle has all but died out. In particular, the construction of the Kara-Kum Canal on the desert's southern fringe enabled the irrigation of extensive cotton fields, and large-scale cattle-raising, as well as industrialization.

The discovery of rich oil and natural-gas reserves have led to further development. Today most of Turkmenistan's population live alongside the canal or in the oases of the Amu Dar'ya.

**In some areas of the Kara-Kum, particularly in the central and northwestern parts of the desert, elements of the nomadic lifestyle persist. The camel, for example, remains a vital form of transportation in those regions with few paved roads or railroads.**

# Fact File

- The Kara-Kum covers a total area of around 300,000 square kilometers (115,830 sq. mi.) and at its greatest extent stretches 800 kilometers (500 mi.) from east to west and 485 kilometers (300 mi.) north to south.

- The name Kara-Kum is also applied to a smaller area of desert to the northeast of the Aral Sea in Kazakhstan, often known as the Aral Kara-Kum.

- The Kara-Kum is an example of a continental desert, in which the near-absence of maritime influences causes aridity. On average the northern part of the desert receives as little as 70 millimeters (2.75 in.) of rain a year, although the south – which is closer to the Arabian Sea – receives more than twice this amount. Most of the rain falls in winter and early spring (December to April).

- The Kara-Kum has a continental-type climate, with long hot summers and cool winters. In the central part of the Kara-Kum average summer temperatures are as high as 34°C (93°F). The north and areas adjacent to the Caspian Sea are less hot. In January average temperatures are 4°C (39°F) in the south and -4°C (25°F) in the north, although temperatures can drop as low as -20°C (-4°F).

- The Kara-Kum was formed some 30 million years ago, when the sea that once covered the region began to recede. As the Amu Dar'ya flowed in various courses over the resulting plain it deposited large amounts of sand and clay, creating the desert.

- Only a small human population lives in the Kara-Kum, with a density of no more than one person every 6.5 square kilometers (2.5 sq. mi.).

- More than 40 different minerals are found in the Kara-Kum's sands, brought down over thousands of years from nearby mountains.

- The hardy, wiry-haired sheep raised in the region are known as karakul, after a village in Uzbekistan. Particularly valued is the soft wool of the karakul lamb.

- Turkmenistan holds the world's fifth-largest reserves of natural gas, with estimates of the country's total gas resource base ranging as high as 15 trillion cubic meters (535 trillion cu. ft.). Much of this resource is found near Mary.

# Iranian Desert

**1. Central Plateau**

The Iranian Desert forms part of the Plateau of Iran, an extensive highland area of western Asia, comprising central and east Iran and western parts of Afghanistan and Pakistan. The region is largely arid and relatively infertile, although the lower slopes of surrounding mountain ranges support some agriculture. The two main arid areas are the salt deserts Dasht-e-Kavīr and Dasht-e-Lūt, both in Iran. Afghanistan also has extensive desert areas.

**2. Dasht-e-Kavīr**

Also known as the Great Salt, or Kavīr, Desert, this vast saline plateau dominates northern Iran and measures approximately 800 kilometers (500 mi.) from east to west and 390 kilometers (240 mi.) from north to south. It is named for its kavīrs, or salt marshes, whose saline crusts are created by the near-rainless climate and intense surface evaporation. The kavīrs act similarly to quicksands and are dangerous to unwary travelers. The area is almost uninhabited, with settlements confined to the surrounding mountain ranges.

**3. Kavīr Buzūrg**

This is the largest of the kavīrs in the Dasht-e-Kavīr, its name meaning simply "Great Kavīr." Measuring some 320 kilometers (200 mi.) long and 160 kilometers (100 mi.) wide, it lies at the heart of the region and is separated from smaller surrounding kavīrs by a ring of sandy hills.

**4. Kavīr National Park**

This national park, which lies only 150 kilometers (93 mi.) from Iran's capital, Tehran, features the mix of desert and arid steppe that characterizes about one-half of the country's land surface.

**5. Tehran**

Iran's capital lies just to the northwest of the Dasht-e-Kavīr, at the foot of the Elburz Mountains. Made the capital of the modern state in about 1788, it developed rapidly in the 20th century and was the center of the Islamic Revolution in 1979.

**6. Dasht-e-Lūt**

Extending south from the Dasht-e-Kavīr is the sand-and-pebble Dasht-e-Lūt, or Lūt Desert, some 480 kilometers (300 mi.) long and 320 kilometers (200 mi.) wide. To the east is a region of sand dunes and sand seas, while to the west are some of the world's most impressive yardangs, or linear rock ridges. In the central depression is the large salt marsh of Namakzār-e Shadā.

**7. Yardangs**

About 160 kilometers (100 mi.) east of the city

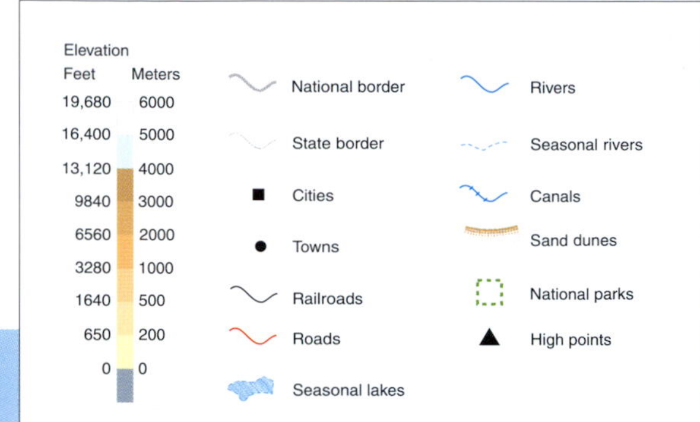


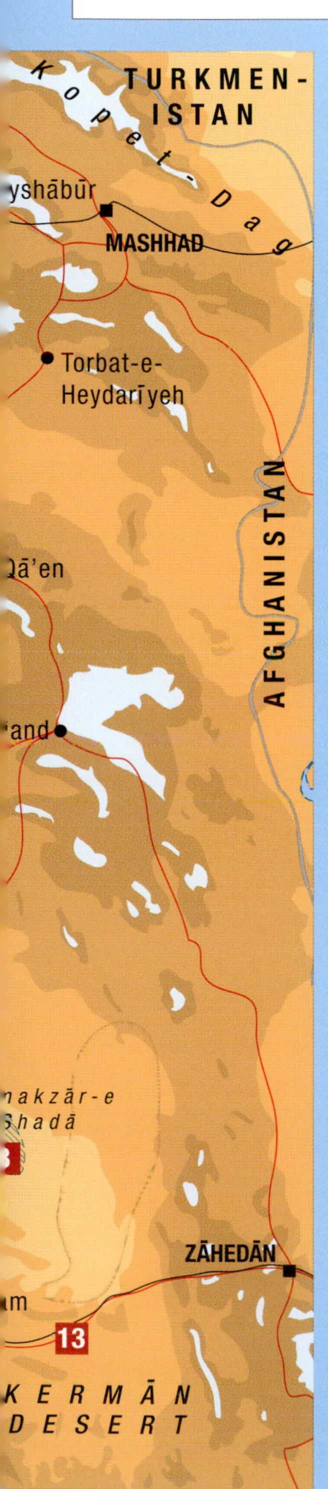


of Kermān are some of the world's most prominent yardangs – long, parallel, sharply crested ridges that are created by prevailing winds in desert regions.

### 8. Namakzār-e Shadā

This salt marsh occupies the Dasht-e-Lūt's lowest depression, less than 300 meters (1,000 ft.) above sea level. It is reputed to have some of the world's highest summer temperatures and lowest levels of humidity.

### 9. Kermān

This city, in ancient times known as Carmana, lies on the edge of the Dasht-e-Lūt. It is a major coal-mining center and also the center of Iran's copper production industry. It is most celebrated, however, for the manufacture of rich shawls and carpets.

### 10. Seasonal lakes

The 442 square kilometer (180 sq. mi.) Daryacheh-ye Tashk (Lake Tashk) is one of the few freshwater lakes in Iran. In the bleak Central Plateau virtually all the bodies of water are highly saline and seasonal.

### 11. Bam

This oasis town, today mainly famous for its beautiful citadel, was in medieval times an important stopover and trading place on trans-Asian caravan routes.

### 12. Oil reserves

Although Iran's oil reserves are concentrated in the southwest of the country, exploration has discovered exploitable reserves elsewhere, including the Dasht-e-Lūt.

### 13. Transportation

Only a handful of roads cross the desert, whose harsh environment discourages transportation or settlement. The Trans-Iranian Railroad links Tehran with the Persian Gulf and the city of Zāhedān close to the Afghan border.

**In this bleak land depression at the heart of the Dasht-e-Lūt in southern Iran is the Namakzār-e Shadā, a vast salt marsh, or kavīr. Kavīrs are characteristic of both the deserts of Iran's Central Plateau and are hazardous to unwary travelers.**

# Fact File

- The Iranian Desert comprises two distinct areas, the Dasht-e Kavīr, or Kavīr Desert, and the Dasht-e-Lūt, or Lūt Desert.
- The Persian word *dasht* means "desert" or "plain."
- Lying a little north of the tropic of Cancer, the Iranian Desert is a high-pressure climate desert (*see* pp. 16–18) whose aridity is maintained largely by the hot, dry air that flows from the tropics. Another contributing factor to the desiccation of the Iranian Plateau is the rain-shadow effect (*see* pp. 18–20) created by the Zagros Mountains in the western part of Iran.
- Rainfall in the southeast of the Iranian Desert is less than 50 millimeters (2 in.), most of which occurs in the winter.
- A characteristic feature of the Iranian Desert is the *kavīr*, or salt marsh or waste, similar to the Arabian sebkha. The hot, dry air of the region rapidly evaporates moisture from land depressions, leaving behind deposits of crystallized salts. These in turn draw in moisture both from the atmosphere and from deep within the earth, drying the air still further and creating a marshy layer beneath the surface.
- From May to September constant strong winds sweep the Iranian Desert, keeping sand in almost constant motion where it is not anchored in place by plant life. Strong winds are also responsible for creating yardangs – wavelike parallel rock ridges – in the western Dasht-e-Lūt.
- Dasht-e-Lūt is known to have rich reserves of oil. Iran is the fourth-largest oil producer in the world, and in the future these largely untapped resources may help maintain Iran's position among the world's leading oil-producing nations.

# Thar Desert

**1. Thar Desert**

Also known as the Great Indian Desert, this vast sandy desert of southeast Pakistan and northwest India stretches from the Indus Plain in the west to the Aravalli Range in the east, and from the Sutlej River in the north to the Rann of Kachchh in the south. Altogether the desert covers some 200,000 square kilometers (77,000 sq. mi.). The desert is a high-pressure (subtropical) climate desert, its aridity caused by the dryness of the monsoon winds that blow across the region. Its name derives from the *t'hul*, the local name for the desert's sand dunes, which may reach up to 150 meters (500 ft.) in height.

**2. Bhakars and dhands**

Among the sand dunes and sandy plains are low barren hills, known as bhakars. Saline lakes, called dhands, are also scattered across the desert.

**3. Climate**

The Thar Desert as a whole has low rainfall, ranging from 100 millimeters (4 in.) the west to 500 millimeters (20 in.) in the east. Summers are very hot – temperatures as high as 50°C (122°F) are not uncommon – and often feature ferocious dust-raising winds, which might have velocities of up to 150 kilometers (93 mi.) per hour.

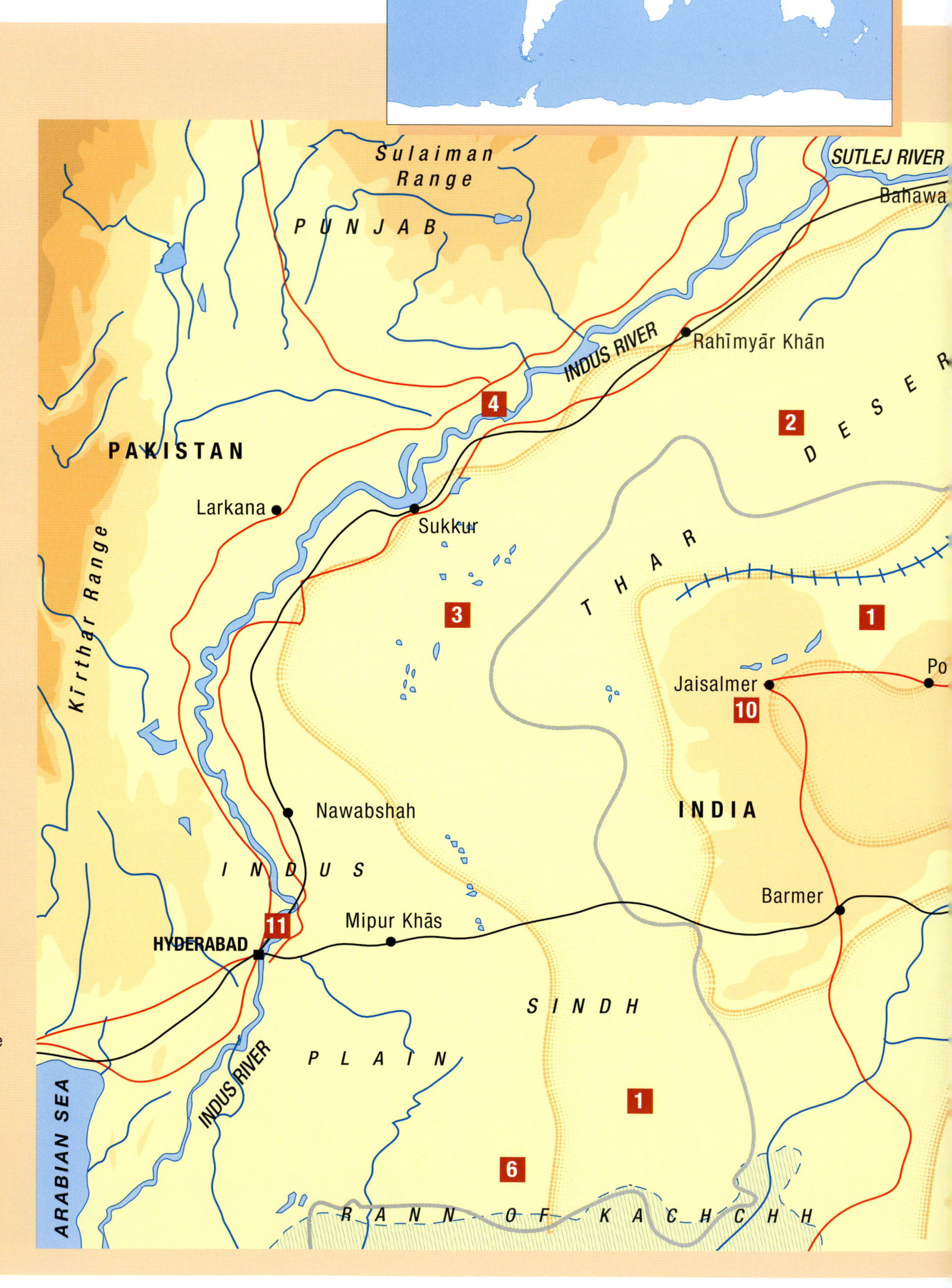

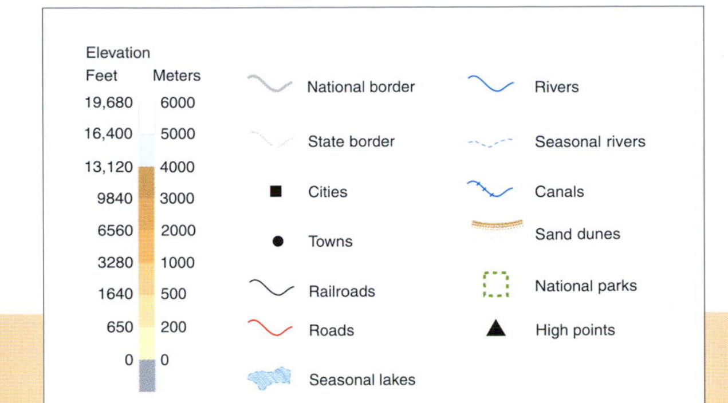


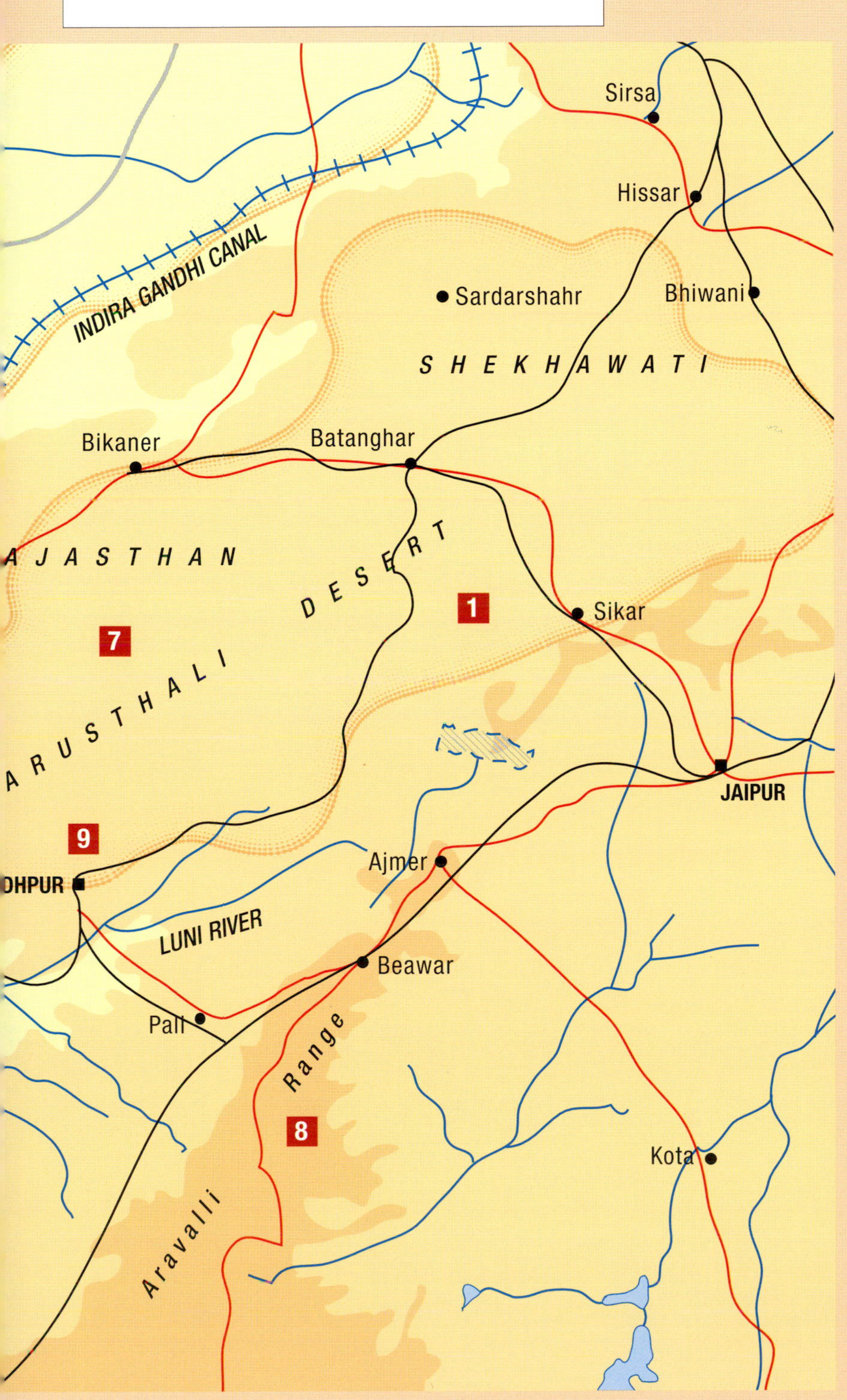


**4. Indus River**
One of the world's longest rivers, the Indus flows some 2,900 kilometers (1,800 mi.) from southeastern Tibet to the Arabian Sea. The river provides the waters for major irrigation projects in the Thar Desert.

**5. Indira Gandhi Canal**
The Indira Gandhi Canal, formerly known as the Rajasthan Canal, carries water 470 kilometers (290 mi.) into the Indian section of the desert from the Harike Barrage on the Indus River. A network of canals also irrigates the Pakistani part of the southern Thar, controlled by the Sukkur, or Lloyd, Barrage. The control of water supplies between the two countries, which are often hostile toward one another, is regulated by the Indus Water Treaty of 1960.

**6. Rann of Kachchh**
These great, salty mudflats cover a total area of around 23,000 square kilometers (9,000 sq. mi.), stretching between the Thar Desert and the Arabian Sea. The region was originally part of the sea but has been filled with sediment and become a huge mudflat. In the monsoon season the plain floods and sometimes turns the Kachchh Peninsula into an island. Settlement in the Rann of Kachchh is only possible on a few hilltops. Because of its aridity and barrenness, the Rann is often considered part of the Thar Desert.

**7. Marusthali Desert**
The eastern portion of the Thar is called the Marusthali Desert, a wasteland of sand dunes whose name means "Land of the Dead." Its southern border is marked by the Luni River, whose waters are often lost beneath the sand. The region's inhabitants raise livestock or grow crops on irrigated land.

**8. Aravalli Range**
These hills cut across Rajasthan, marking the eastern border of the Thar Desert. Beyond the range is the state's more productive and fertile southeast region.

**9. Jodhpur**
This walled city on the southern edge of the Marusthali Desert, founded in 1459, was the capital of the former state of Jodhpur. The Jodhpur region was famed for its camel breeding and gave its name to a type of riding breeches.

**10. Jaisalmer**
A major caravan center since the 12th century, Jaisalmer is a marketplace for camels, hides, wool, and other goods.

**11. Hyderabad**
The Pakistani city of Hyderabad is an important rail center and is also known for its skilled metalworkers.

# Taklimakan

### 1. Taklimakan

This desert of shifting sand dunes occupies some 323,750 square kilometers (125,000 sq. mi.) in the Tarim Basin, western China. It is a continental desert; that is, its aridity is largely caused by its sheer distance from the sea. An additional causal factor, however, lies in the rain-shadows created by the towering mountain ranges that border the Taklimakan on three sides – the Tian Shan on the north, the Kunlun Shan on the south, and the Pamirs to the west. Some parts of the desert receive only 10 millimeters (0.4 in.) of rainfall a year, and nowhere gets more than 38 millimeters (1.5 in.).

### 2. Sand dunes

The Taklimakan is one of the world's largest sand deserts, and its windblown sands may be up to 300 meters (1,000 ft.) thick. The complex winds of the region have created a great diversity of sand formations, including longitudinal, transverse, and star dunes. The eastern foothills of the Tian Shan mountains contain some of the world's largest sand dunes, which may reach heights of 200–300 meters (650–1,000 ft.).

### 3. Rivers

The main river of the Taklimakan is the Tarim, which, formed by the waters of the Yarkant and Hotan, flows west–east along the desert's northern edge. Most other rivers in the region flow roughly north off the Kunlun Shan, but extend at best only for about 100–200 kilometers (60–125 mi.) before becoming submerged in the desert sands.

### 4. Temperature

The desert's climate is continental, with hot summers and cold winters and with a temperature range of 21°C (70°F). On the eastern edge of the Taklimakan summer temperatures may reach as high as 38°C (100°F).

### 5. Turpan Depression

This depression in the far northeast of the Tarim Basin is sometimes included as part of the Gobi Desert and has numerous salt lakes. Its lowest point is 130 meters (426 ft.) below sea level. The highest temperature recorded in China, 48°C (118°F), was at the town of Turpan.

### 6. Lop Nur

This arid salt marsh receives the Tarim River but has no outflow. This desolate area has been the test site of China's nuclear weapons since 1964.

### 7. Qiemo

This town is one of several sites in the Taklimakan where archaeologists have discovered remains of some of the region's ancient inhabitants, some dating back to 2000 B.C.

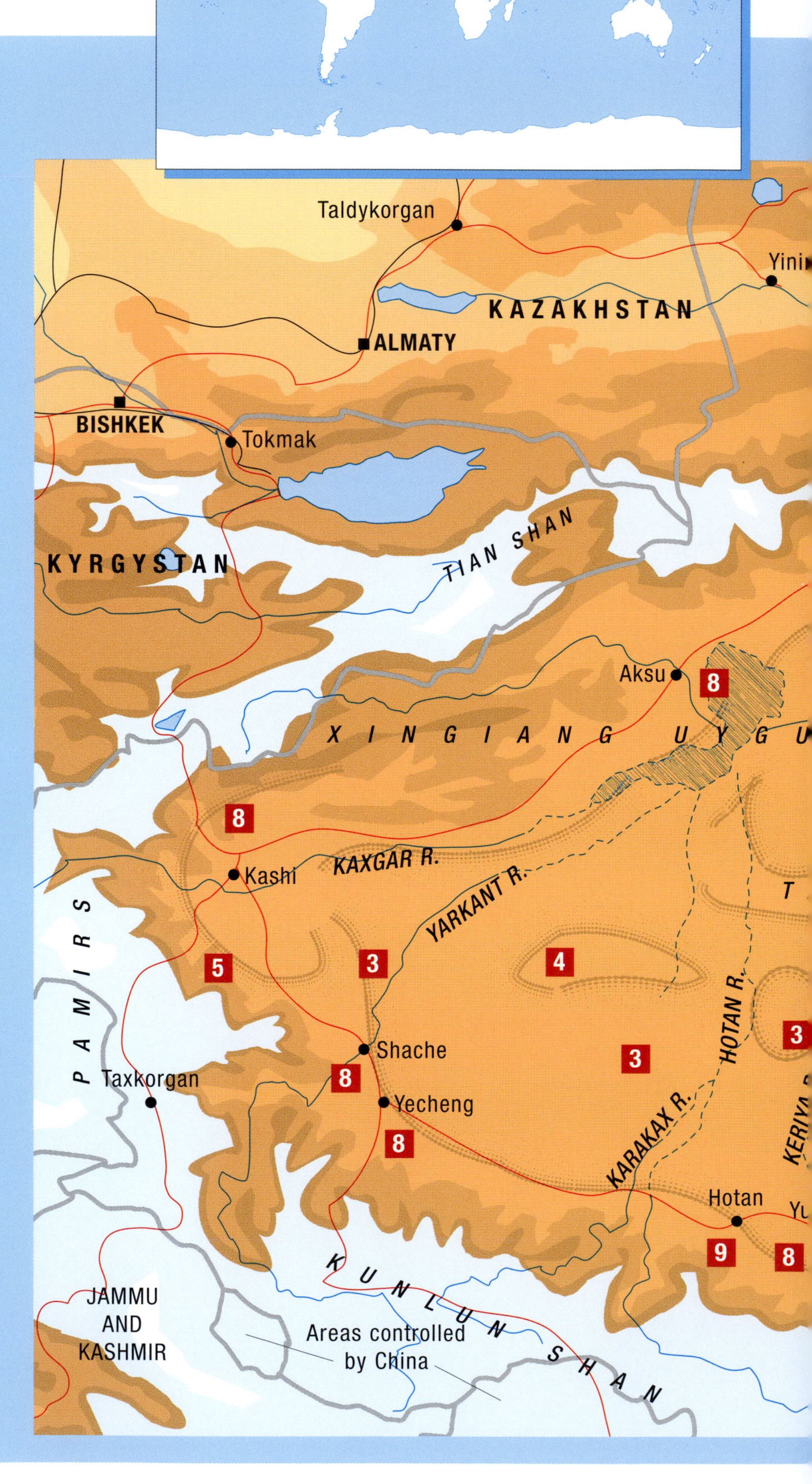

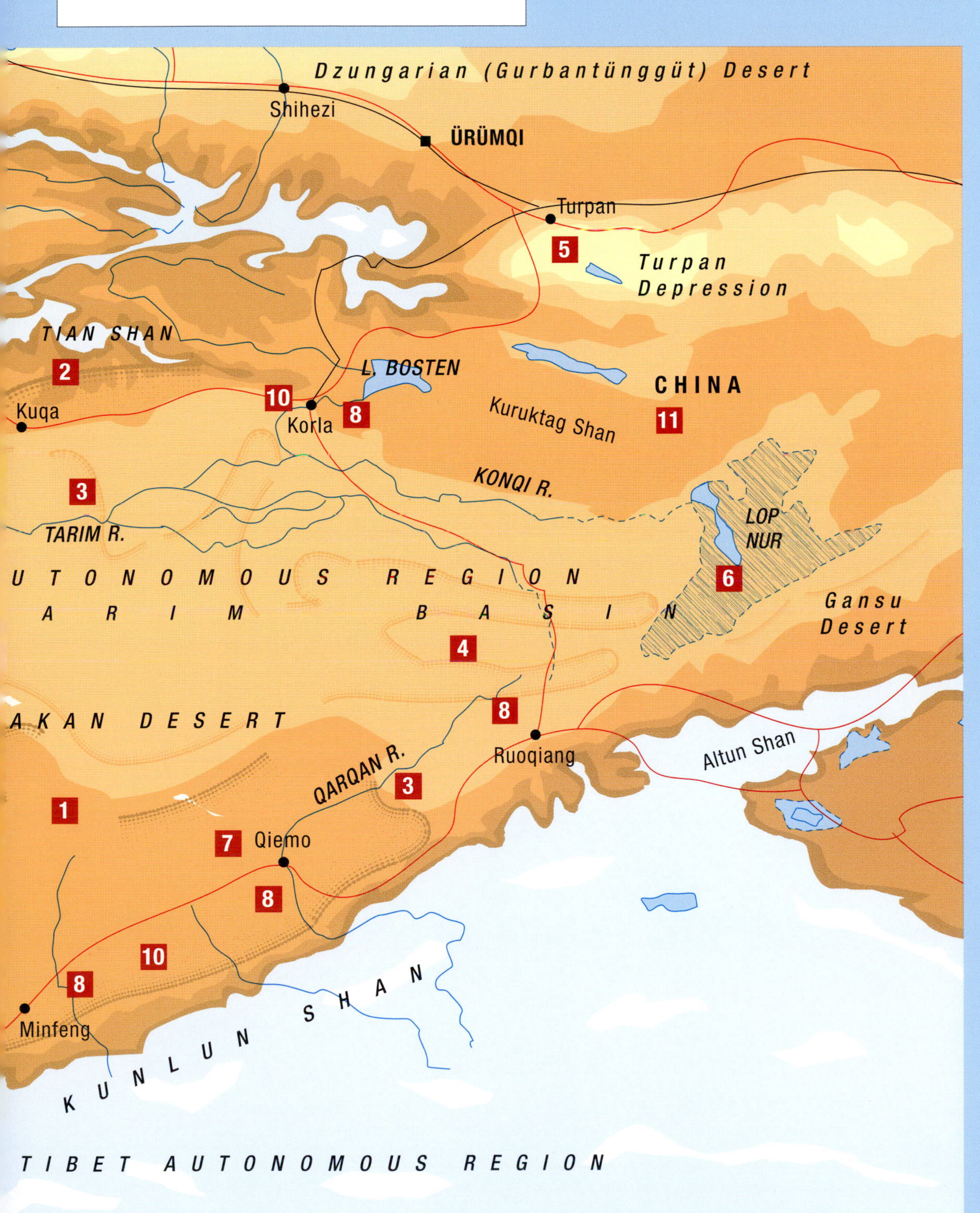


**8. The Silk Road**

The historic caravan routes from China to Central Asia and Europe ran along the northern and southern edges of the Taklimakan (*see map*, p. 138). Oases such as Hotan, Aksu, and Kuqa became important supply bases.

**9. Hotan**

This was the most important stopover on the Silk Route on the desert's southern edge. The town became a flourishing center of Buddhist culture, which was introduced to the region from India along the caravan routes. During the late 19th and early 20th centuries explorers uncovered Buddhist monasteries and artifacts that had long been buried deep under the shifting sand.

**10. Oil reserves**

In the 1950s vast reserves of petroleum were identified in both the northern and southern extremities of the Taklimakan. To date, most of the drilling work has been carried out by the Chinese government around the town of Korla.

**11. Desertification**

Much of the semiarid grassland fringes of the Taklimakan Desert is threatened by desertification. Overgrazing and poor farming have rapidly exhausted the already low fertility of the land and turned former grasslands into desert.

# Gobi Desert

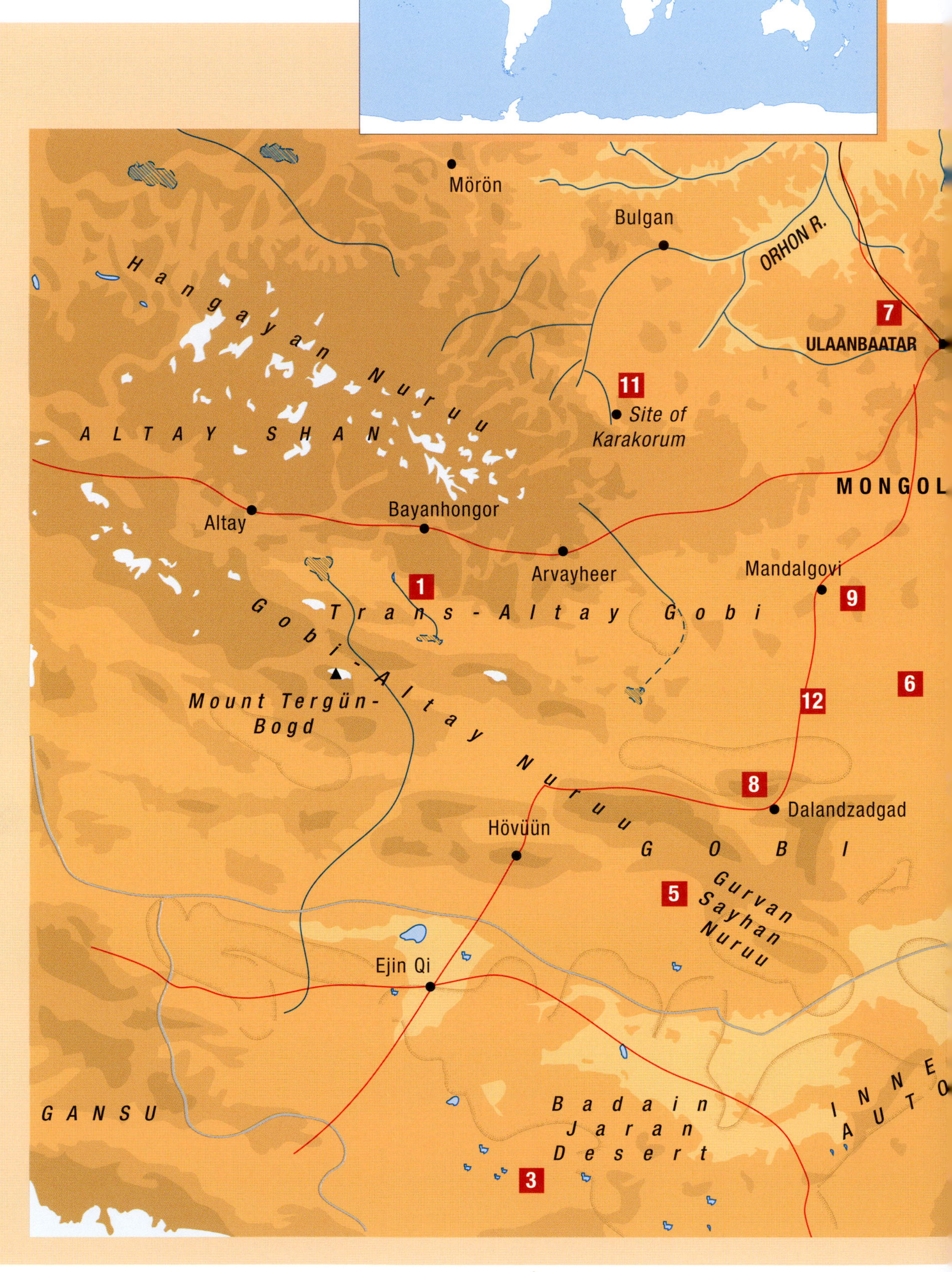


**1. Trans-Altay Gobi**
This rugged, desolate region of the western Gobi is made up of broken plains, high mountain ranges, dry gullies, and salt marshes. It is extremely dry, receiving less than 100 millimeters (4 in.) of rain per year; it is sparsely vegetated.

**2. Eastern Gobi**
The eastern part of the Gobi is wetter than the western, with up to 200 millimeters (8 in.) of rain per year. Although there are few rivers, there is plentiful underground water, which breaks the surface as small lakes and springs. Despite this, the region is only thinly vegetated.

**3. Badain Jaran**
The southwest region of the Gobi is almost entirely sand, and its southerly parts are known as the Badain Jaran Desert. The star dunes here are up to 500 meters (1,640 ft.) tall. Still further south is another sandy desert, Tengger Shamo.

**4. Mu Us Desert**
Not strictly part of the Gobi, this sandy desert plateau, otherwise known as the Ordos, fills the great northern bend of the Huang, or Yellow, River. In the 1960s the Chinese government planted drought-resistant trees along the desert's eastern and southern edges to prevent its sands spreading.

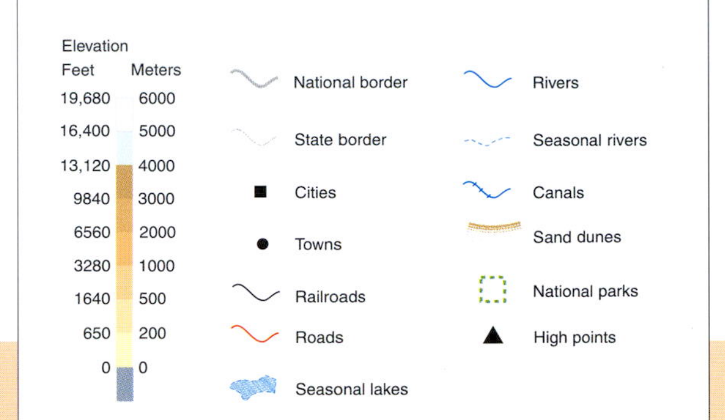


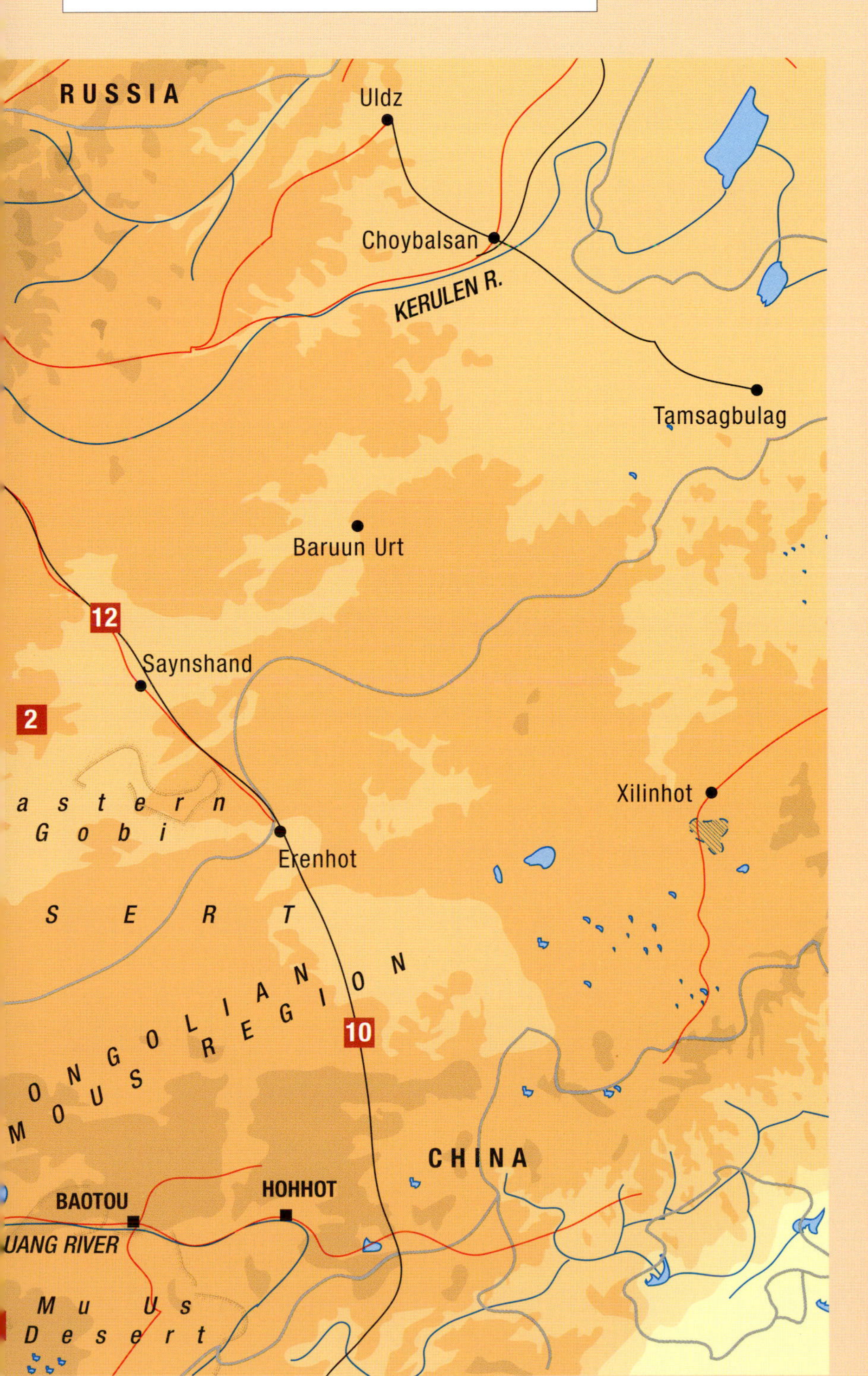


**5. Gurvan Sayhan**

These mountain ranges in the southern Gobi are home to the beautiful Yelyn Valley National Park, which has deep gorges surrounded by towering craggy cliffs where condors make their nests.

**6. Rivers and streams**

Mountain streams are found only on the Gobi's fringes and quickly dry up as they disappear into the desert's loose soils or salt marshes. Underground water is widespread and is sufficiently mineral-free to permit cattle-raising – a mainstay of the Mongolian economy.

**7. Ulaanbaatar**

On the northern fringes of the Gobi is the Mongolian capital, a city of some 550,000 inhabitants. Formerly named Urga, the city was founded in the 17th century as a center of Mongolian Buddhism and later became an important stopover on caravan routes between China and Russia. Its present name, meaning "Red Hero," was adopted after the country became a people's republic in 1924.

**8. Dalandzadgad**

A little over 10,000 people live in Dalandzadgad, some 515 kilometers (320 mi.) by road from the Mongolian capital. They work mining brown and bituminous coal and also produce cement.

**9. Mandalgovi**

This Mongolian town is on the northern transitional grassland zone of the Gobi, about 300 kilometers (186 mi.) south of Ulaanbaatar. Owing to the area's aridity and the harsh terrain, the town's economy is dominated by pastoralism.

**10. Great Railroad**

The Trans-Mongolian Railroad links the Mongolian capital, Ulaanbaatar, with the Trans-Siberian Railroad at Ulan-Ude in Russia and with Erenhot in China. The railroad was built in the mid-1950s, during a period of close Sino-Soviet cooperation, and provides the fastest railroad link between Beijing and Moscow.

**11. Karakorum**

On the headwaters of the Orhon River, on the northern borders of the Gobi, stand the ruins of Karakorum. The Mongol leader Genghis Khan founded a city here (on the ruins of an earlier, Uighur one) in the 13th century, using it as a base to build a huge Eurasian empire. The city was finally abandoned in the 16th century.

**12. Road, air, camel**

A few highways cross the western and eastern Gobi, particularly linking Ulaanbaatar with outlying towns. Air travel is well developed, too. However, in more remote areas of the Gobi camels and horses remain vital methods of transportation.

## Fact File

- The Gobi covers an area of around 1.3 million square kilometers (500,000 sq. mi.) – not much smaller than the largest U.S. state, Alaska. This desert and semidesert area roughly stretches in a great east–west arc – concave to the north – that extends through Central and eastern Asia, covering portions of Mongolia and China.

- The Gobi takes its name from a Mongolian word meaning "waterless place" or "desert." There are no surface rivers in the Gobi proper, although there are underground water sources, which support the region's widespread pastoralism.

- Some authorities include eastern areas of the Tarim Basin in the Gobi, including the Gansu and Dzungarian deserts (*see map*, pp. 68–69). Other, smaller desert areas to the south of the Gobi are also sometimes treated separately. These include the Mu Us, or Ordos, Desert and the Tengger Shamo (Ala Shan).

- The Gobi is in large part bare rock, with areas of sparsely vegetated dry brown or gray soils. Only the southwest part has extensive areas of sand dunes.

- The annual rainfall in the northeast is more than double that in the west – 20 centimeters (8 in.) versus 69 millimeters (2.75 in.).

- The Gobi supports fewer than one person per square kilometer (3 people per sq. mi.).

# The Gobi

The Gobi is one of the world's great deserts. For hundreds of years after it was first described by the Venetian traveler Marco Polo (*see* p. 140) in the 13th century it existed in the European imagination as one of the last unexplored regions of the world, a forbidding and inhospitable vastness of rock and sand. Parts of the Gobi are indeed barren and harsh – with little rainfall, bitter winters, and long hot summers – but others are relatively benign. In the east the desert experiences what is virtually a monsoon climate, with regular dry and rainy seasons; in the northeast the landscape and vegetation come to resemble those of the great Asian grasslands, the steppes. The desert has been home to cattle-herding Mongolian nomads and their ancestors for millennia.

There is relatively little vegetation in the Gobi proper. Much of what there is occurs as shrub or bushes on the flatlands: there are no trees. There is some grass, however, of the variety *echinochloa*, and on the salt marshes created by the Gobi's underground water supplies the potash bush, the Siberian nitre bush, and the tamarisk grow. The sandy regions support typical Asian desert plants, including saxaul and wormwood. At the edges of the desert plains, however, the story is rather different. Grass flourishes – desert grass on the steppes and the lower slopes of the mountain ranges, and Gobi feather grass and other feathery varieties higher up the mountain slopes. Herbaceous plants often grow in great meadows among the grassland, including rhizomes – elongated roots that give off only small shoots – called Mongolian onions.

Aside from being home to the usual array of reptiles and insects, the Gobi also supports a few larger mammals living in the wild. Wild camels are common, as are kulans (*Equus hemionus hemionus*), hardy Asiatic wild asses that roam in large herds during the

**This encampment of Mongolian pastoralists in the Gobi features the traditional cone-shaped yurts or *gers*. For ease of transportation, the interiors of yurts are generally simple. However, more permanent examples are often quite lavish, cosy affairs (*see illus.*, p. 154).**

**The typical Gobi landscape features gray, sparsely vegetated plains, with low-lying, eroded hills and limpid blue skies.**

winter season, and the dzeren antelope. The most famous mammal of the Gobi is Przewalski's horse (*Equus caballus przewalskii*) – the last wild horse subspecies to survive into the 20th century – although it now seems to be extinct, except in a few protected environments. This ponylike horse gets its name from the Russian explorer N.M. Przhevalsky, who first discovered it in western Mongolia at the end of the 19th century.

The Gobi is very thinly populated, except on its lusher, semidesert fringes and around river valleys, where there is also rich evidence of the region's ancient and medieval civilizations.The rural Mongolian population (*see* pp. 139–140) of the Gobi remains largely nomadic, migrating with their herds of sheep, goats, cattle, Bactrian camels, and, to a lesser extent, horses. Rare among desert peoples, and despite the turmoils of the 20th century, the Mongolian nomads have maintained much of their traditional lifestyle – typified by the cone-shaped, latticed tents known as a yurts or *gers*. In many areas these striking white structures still dot the gray and green landscape. Settled populations on the Gobi's borders – of both Mongolian and Chinese ethnicities – are usually agriculturalists; a small percentage work in salt, coal, and metal mines.

The Gobi has few roads, and most of these are unpaved; many of the ancient caravan routes between Russia and China are still in use. The Trans-Mongolian Railroad crosses the desert landscape from Ulaanbaatar to Erenhot.

# PLANTS OF THE DESERT

*Plants need water to survive, but even in the desert they have been able to adapt their shape, leaves, and metabolism in the struggle to conserve moisture. As much as 90 percent of the mass of some desert-dwelling plants lies below the surface in water-retaining root systems.*

Drought can occasionally affect plant species in almost every biome, even rain forest, and everywhere plants have made adaptations that minimize water loss. The needlelike leaves of conifers, for example, have evolved to minimize water loss by transpiration (evaporation into the air from exposed plant tissues). Similarly, plants growing in sandy, coastal areas have modified to hold moisture in a salt-ridden environment that tends to draw water from them.

Adaptation is crucial in the desert, where drought is the norm rather than the exception and where plants are daily afflicted by the desiccating effects of the hot sun. Some of the adaptations found in arid regions – such as succulent leaves and thick protective surfaces – are similar to those found in coastal plants. Other desert adaptations are unique. The characteristic desert plant is the cactus, and its adaptations, including fleshy stems and scales or spines instead of leaves, make it an efficient drought survivor.

**Pink sand verbena (*Abronia villosa*) and white dune evening primrose (*Oenothera deltoides*) bloom in the desert of southern California. Such floral displays occur briefly in deserts all over the world in the wake of heavy rains.**

Other species have adopted even more spectacular strategies. In the aftermath of rain, deserts can be briefly transformed from barren wildernesses into extravagant paradises, smothered with a tangle of brilliantly colored flowers – poppies, tulips, and desert peas, among others. In the main, however, plants of the desert survive by being very unspectacular. They grow unobtrusively close to, or even beneath the ground, with few showy leaves and flowers. Their root system is often their most developed part.

The successful adaptation of plants to one of the planet's harshest environments is the foundation for all other life in the desert, providing the first and most vital link in its food chains. The importance of plant life in every environment is nowhere clearer than in marginal, semiarid areas, where its loss or degradation is the principal cause of desertification (*see* Chapter 8).

### Adapting to Deserts

In simple terms a plant acts as a conduit for water from soil to air, drawing in moisture through its roots, extracting the nutrients it requires to grow, and finally giving off vapor into the air from its

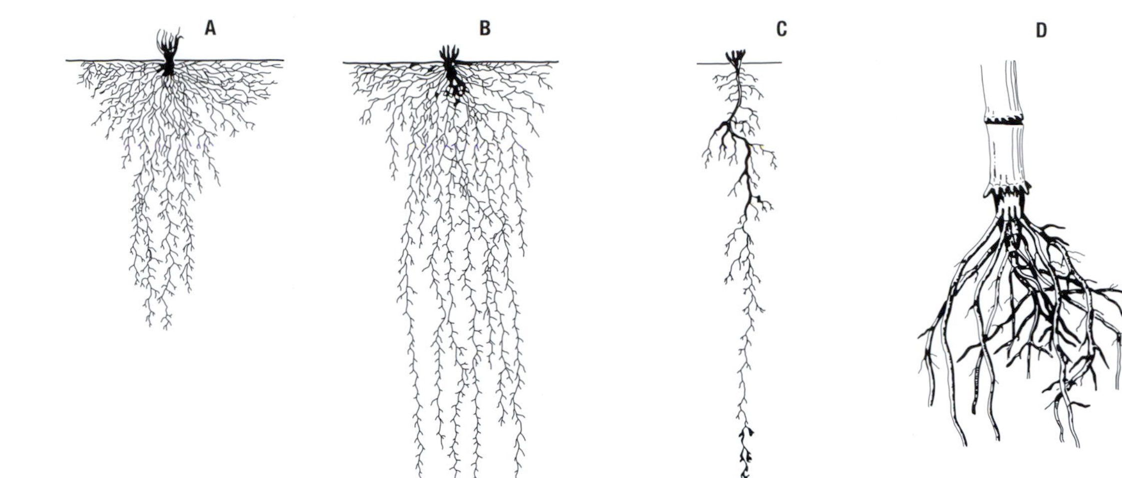


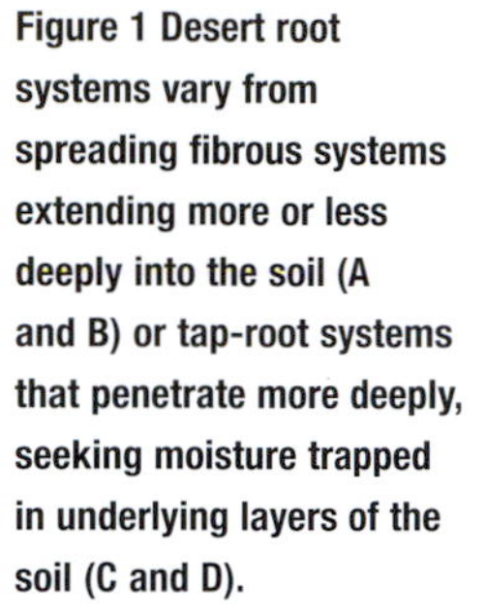
**Figure 1 Desert root systems vary from spreading fibrous systems extending more or less deeply into the soil (A and B) or tap-root systems that penetrate more deeply, seeking moisture trapped in underlying layers of the soil (C and D).**

leaves. An organic system of this kind would ordinarily be unsuited to desert life, where water is rare and elevated daytime temperatures evaporate water directly from the plant's exposed surfaces, scorching the plant's outer tissues in the vegetable equivalent of sunburn. It might be expected, therefore, that deserts would be plant-free wildernesses of bare rock and sand. That this is not always the case is a tribute to the remarkable adaptive power of plants as they have evolved to cope with drought, water loss by transpiration and evaporation, sun and wind exposure, high and low temperatures, and the ravages of grazing herbivores.

## Finding water

In some desert environments, such as the shifting dunes of the Rub' al-Khali, or Empty Quarter, in the southern Arabian Peninsula (*see* pp. 40–41), it is

## PRIMARY PRODUCTIVITY

Biologists and ecologists measure and compare habitats using various standard features. One such feature is *primary productivity* – stated simply, the weight of new plant growth, including matter above, on, and in the soil, each year in a certain area, normally taken as one square meter. As might be expected, the desert habitat scores low in a table comparing the average productivity of major habitat types. The low annual productivity of deserts results directly in the sparsity of their animal life, for all the resourcefulness of its adaptations. By contrast, the richly productive tropical rain forests are home to a seemingly endless proliferation of zoological species.

Another way of measuring plant life – and one that is more revealing when looking at the desert habitat – is the amount of *standing biomass*, which aims to give a "snapshot" measurement of the living weight of plant matter at any one time (although in practice, it is sampled and averaged over many years). In the world's most productive habitats, such as tropical rain forests, biomass varies from about 40 to more than 80 kilograms per square meter. Most of this is above ground in the form of trunks and branches of trees, since their root systems do not need to be extensive. Temperate deciduous forests also show a high level of biomass, some 40 to 50 kilograms per square meter, while grasslands by contrast have a surprisingly low value – rarely above 5 to 7 kilograms per square meter – owing to the lightweight nature of grass itself. Using the standing biomass system, different deserts score very widely, from 1 to 20 kilograms per square meter. However, some 80 to 90 percent of this is below ground, as roots, bulbs, tubers, corms, and other underground plant parts. It is for this reason that burrowing herbivorous creatures make up a large proportion of a desert's animal life and that deserts look so empty on the surface.

COMPARATIVE PRODUCTIVITY OF MAJOR BIOMES

(Average annual productivity in kilograms per square meter per year)

|  |  |
|---|---|
| Tropical rain forest | 2.5 (range 2–3.5) |
| Tropical swamp | 2.5 (range 2–3) |
| Temperate rain forest | 1.2 (range 1–1.5) |
| Temperate deciduous forest | 1.0 (range 0.5–2.5) |
| Tropical grassland | 1.0 (range 0.2–2) |
| Dry scrub | 0.8 |
| Northern conifer forest | 0.5 (range 0.2–1.5) |
| Semidesert | 0.2 |
| Desert | 0.05 |

A creosote bush of the species ***Larrea tridentata*** sends out shallow roots into the dry, heavily eroded sandstone of Nevada's Valley of Fire. In some areas of arid southwest America this shrub forms a dense scrub.

true that no plants can grow. In addition to the burdens of heat and drought, the movement of windblown sand prevents any seeds from taking root. In most other deserts, however, at least a few hardy plants survive, optimizing available water through three main adaptations, used separately or in combination: extensive or deep root systems that gather as much moisture as possible from the surrounding soil; body parts that are able to store water securely for considerable periods, and features that reduce transpiration from exposed leaves and stems. Botanists term plants that are well-adapted to dryness and drought xerophytes, that is, "dry-loving."

Desert plants have some of the longest roots of any members of the plant kingdom (*see* Fig. 1 *opposite*). Those of the North American desert mesquites (*Prosopis* species), for example, descend almost vertically for up to 15 meters (50 ft.), seeking out deep subsoils that remain permanently moist, while the climbing Queen of the Night cactus (*Peniocereus greggi*) has a single vertical tap root that can weigh more than 30 kilograms (66 lb.). In Africa the raddiana acacia (*Acacia raddiana*), which often grows in desert wadis, or dry riverbeds, also sends roots many meters deep.

The roots of some other desert plants spread outward rather than downward. Many cacti and euphorbias follow this tactic, extending a fine network of thready roots into the surrounding soil like an upside-down umbrella, channeling rainfall to the plant near the center. The creosote bush (*Larrea tridentata* and others) of southwest North America also has fairly superficial spreading roots, which not only gather water but also help to propagate a single plant into a set of surrounding clones over hundreds of years.

Roots are not necessarily the only water-gathering parts of a plant. Some arid-area species, including the pygmy cedar tree (*Peucephyllum schotti*) and the welwitschia (*Welwitschia mirabilis*; *see* p. 84), absorb water as dew or condensed mist through their surfaces. Some cacti and other spiky or thorny plants have downward-pointing spines that concentrate rainwater into drips that roll onto the ground below, ready for absorption by the roots.

### Holding and keeping water

Once obtained, valuable water must be retained. Some trees of semidesert regions have evolved greatly swollen trunks that act as living water butts. One of the most distinctive of these is the baobab (*Adansonia digitata*) of Africa. The squat, bottlelike trunk of mature specimens regularly approaches 10 meters (33 ft.) in diameter and with reputed circumferences exceeding 50 meters (164 ft.). The wood within the trunk and lower branches is soft and spongy, and can hold vast quantities of moisture. During the dry season the baobab sheds its leaves to reduce transpiration and relies on stored water, sometimes shrinking visibly as it draws on its water supply.

The best-known water-storers are the cacti, which, along with the agaves and euphorbias are termed succulents (meaning "juicy"). Cacti's stems can swell like balloons as they suck up water from the soil. The stem surfaces are often folded or pleated so that as the stem expands the plant is able to increase its internal volume without having to produce additional surface tissues. The storage capacity of some of the larger cacti is remarkable. For example, the organ-pipe cactus (*Lemaireocereus thurberi*), which can grow up to 6 meters (20 ft.) tall, is able to absorb more than 500 liters (875 pt.) of water in the days immediately following a rainfall – a reservoir that will support it for up to four months of drought.

**The thick, fleshy leaves of the agave, a genus found in arid and semiarid regions of the Americas, are excellent water-storers. The thorny, serrated edges serve to ward off hungry and thirsty herbivores. The coastal agave (*Agave shawii*) shown here is found in the arid Baja California peninsula.**

## Reducing water loss

Cacti are excellent examples of the strategies employed by plants to reduce water loss. Typically, a plant loses moisture mainly through the surfaces of its leaves – especially through the tiny holes called stomata that are scattered chiefly on the undersides of the leaves. During the process of photosynthesis, a plant absorbs carbon dioxide from the atmosphere and liberates oxygen into the air: the stomata enable this exchange, absorbing carbon dioxide and leaking water vapor.

Many desert plants have adapted simply by having fewer or smaller stomata on their leaves compared to species living in more equable habitats. Some species, however, have evolved more drastic solutions. The mulga (*Acacia aneura*), a dry-adapted acacia that forms the scattered woodland of large regions of outback Australia, is a small tree only 5 to 10 meters (16–32 ft.) tall, with spreading gray foliage and thin seedpods. What look like its leaves are actually the leaf stalks, or petioles, which have become enlarged and flattened into tough, almost thorny strips, called phyllodes. Typically, leaf stalks have far fewer, if any, stomata per unit surface area than the leaves themselves, and, in the case of the mulga, it is the green phyllodes that have taken over the photosynthesizing function, thus minimizing water loss. The strong, leathery coverings, or cuticles, of the mulga's phyllodes also cut down evaporation. Similar cutin- or wax-covered surfaces are a feature of many desert plants, notably cacti.

Cacti have also taken the evolutionary theme of leaf reduction one stage further. Many species of the family have spine- or thorn-shaped leaves that fulfil the double role of cutting down transpiration and furnishing protection against herbivores. In this instance, the green tissues that enable photosynthesis to take place are in the stem itself, which compared to a leaf – like the phyllodes mentioned above – has far fewer, if any, stomata per unit area.

## CAM TO THE RESCUE

Because they have no way of storing carbon dioxide, most plants must keep their stomata open during daylight in order to obtain an ongoing supply. Cacti and many other succulents, however, are able to store this vital gas through a biochemical process called crassulacean acid metabolism, or CAM, photosynthesis – so known because it was first described in the plant family *Crassulacaea*, which includes stonecrops and jades.

In the case of the cactus photosynthesis takes place in the green outer cells of its prominent stems. The stomata here open at night, drawing carbon dioxide into the cells, where it combines with PEP (phosphoenolpyruvate) to produce oxaloacetic acid, easily stored within the plant as malic acid. With the arrival of dawn, the stomata close to prevent the loss of water vapor in the rising heat, while the malic acid is broken down to release carbon dioxide for the daily process of photosynthesis.

## Creating microclimates

Plants, unlike animals, cannot hide from the desert's environmental extremes – burning sun, freezing winds, sandstorms, and flash floods – and must make adaptations to survive these perilous conditions. The adoption of light colors, helping to reflect the sun's rays and so weaken their heating effects, is a common example. The branches of the quiver tree (*Aloe dichotoma*; *see illus. opposite*) of the deserts of southern Africa, for example, have a coating of chalky-white powder for this purpose, while the pale blue-green foliage of the blue mallee (*Eucalyptus gamophylla*) of Australia works to the same effect, the thick, waxy leaves also acting as reservoirs of water.

The ribbed or pleated shapes of many cacti, such as the browningias of the Sechura and

Atacama deserts in South America, cast moving shadows on themselves as the sun passes across the sky. This self-created, temporary shade helps to keep the plant's surface area cooler compared to the full glare that would be suffered by a smooth, rounded, naked stem.

## Repelling hungry herbivores

Rare desert plant life is vulnerable to grazing and browsing animals and numerous plants have evolved physical and chemical strategies in order to fend off their ravages. The physical strategies generally comprise defensive features such as spines, prickles, and thorns. Many New World cacti have their leaves modified as sharp, needlelike thorns, which form prickly clusters that spring from specialized lateral buds or side shoots on the stem, called areoles. A unique characteristic of the cacti, areoles also give rise to hairs, barbed bristles known as glochids, and often beautiful flowers and blooms. It is perhaps helpful to think of the cactus as a green trunk, with the areoles as its miniature, condensed branches.

Similar to cacti and almost as spiny are the euphorbias (genus *Euphorbia*) native to Africa, which include the almost treelike ammak (*Euphorbia ammak*) found in rocky areas of the southern Arabian Desert. Unlike the cacti, however, euphorbias' spines are generally borne in pairs rather than clusters, and they also have a different

**The striking, skyward-pointing branches of the quiver tree (*Aloe dichotoma*), or kokerboom, of the Namib and Kalahari deserts, southern Africa, are coated with a white powder that reflects harmful heat rays. The tree's surmounting clusters of daggerlike leaves also help to provide much-needed shade.**

flower structure. The vicious spines of acacias have earned them the common name of thorn trees.

Chemical defense strategies come in the form of nasty-tasting or toxic substances found in the plant's sap, stems, or bark. One of the best-known is the creosote bush (*see illus.*, p. 76) of the North American Southwest. Its waxy leaves contain the poisonous scented oil that gives the bush its common name. Very few creatures risk eating the leaves, although the creosote-bush grasshopper survives almost nowhere else. There are also desert-adapted members of the onion family (genus *Allium*) that produce a stinging sap. More than 150 species of wild cousins of the onion are found in the arid lands from the Middle East across to the Gobi Desert.

## DOUBLE DEFENSES

Throughout the natural world, cooperation as well as competition exists between species, and this is no less true of the desert, where scarcity of resources accentuates the importance of both these survival strategies. Cooperation, or symbiosis, can occur between very different living beings, as can be seen in the complex partnership that has developed between certain ants and acacias. The ants nibble at the bases of the acacia's thorns, provoking the tree to produce hard, protective growths, called galls (*see illus. above*). These enlarge to the size of a grape or apple, and the ants then hollow them out to create living quarters for their nesting colonies. Not only does the tree provide the ants with a safe, spiky place to live; it also provides them with a rich source of food, in the form of its nectar and a protein-rich nutrient produced from special growths on its leaflets, called beltians. In return, the insects keep the tree clean and clear of dust and other debris, even biting through creepers and vines that try to grow up its trunk. They bite and sting leaf-searching herbivores large and small, swarming in seconds to the site of an attack.

### When the desert blooms

Despite the multitudinous adaptations to desert life, plants actually follow one of two basic life strategies. Many of the species mentioned above, including aloes, euphorbias, numerous cacti and other succulents, and trees and bushes, are perennial, staying put for several years – not always growing, it is true, yet ready to thrive when conditions are suitable.

A contrasting strategy to this permanence is ephemerality. Ephemerals grow, flower, and set seed all within a few days or weeks, usually after rain (*see illus.*, p. 74). Then, as the sun comes out and the landscape becomes parched once more, the plants shrivel and die, leaving behind tough-cased seeds that will lie dormant through the long drought to come. There are innumerable species of desert ephemerals, but their life cycle can be illustrated by one example. Sturt's desert pea (*Clianthus formosus*), a native plant of Australia and floral emblem of the state of South Australia, was named after the English explorer Charles Sturt, who observed its brief blooming near Broken Hill, New South Wales, in about 1855.

**After rare rainfall, the scarlet flowers of Sturt's desert pea (*Clianthus formosus*) carpet whole areas of Australia's outback.**

In the Australian outback drought can persist over several years, during which the seeds of the desert pea lie dormant. After significant rainfall, however, the seeds rapidly revitalize, producing stems that twist and scramble a meter or two across the scorching desert surface. The plant grows as a low, trailing herb, with finely divided, grayish-green stems and leaves, both with hairy coverings that reduce evaporation of water from their surfaces. Roots finger their way quickly into the still-damp soil, lengthening by as much as several centimeters daily. Soon some of the stems take a right angle and grow upward to produce the spearlike, bipartite, glossy red flowers, each with a purplish-black "eye spot" at its reflexed center. Occasional plants bearing red and white petals add to the brilliant, mosaic effect.

Within two or three weeks of rain swathes of the desert pea drift over the outback's red soil, the flowers ripening to pods containing small black seeds. Then, as the dryness takes hold again, the pods split or are eaten by animals and the seeds scatter, able to withstand at least five years' lack of moisture before bursting again – phoenixlike – into life and color again.

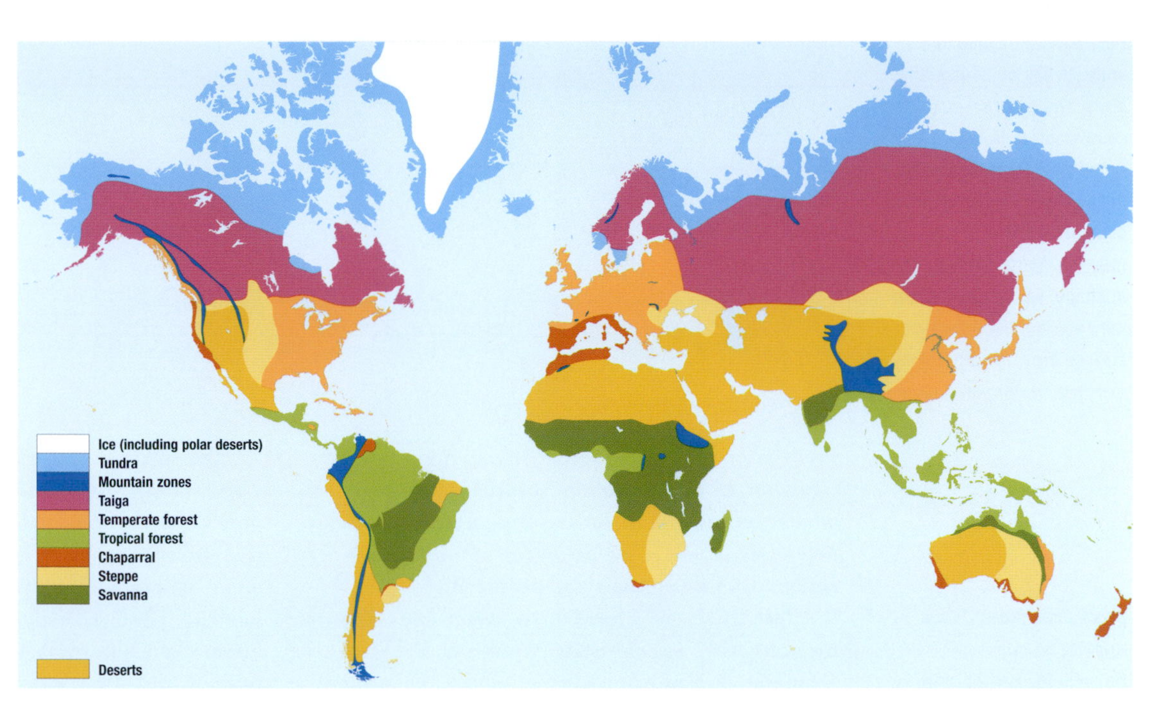


**Figure 2** This map shows the world's major biomes, or major habitat types. The world's deserts and semiarid lands are home to a unique plant life, many of which are dry-adapted relatives of species found in other biomes, but some of which may be the descendants of the plants that grew in ancient deserts, such as the welwitschia of the Namib.

## A Desert Herbal

Though all are characterized by low precipitation and rapid evaporation of moisture, the world's deserts vary greatly in ways that determine their flora. Elevation, average temperature, saltiness, soil type, and the presence of underground water can all play a decisive part in dictating what plants flourish.

Shrubs – particularly sagebrushes – dominate the colder deserts of inland Asia and the mountain zones of North America, and thawing winter snow stimulates a flowering of annuals in spring. Predominant in the tropical deserts and semiarid scrublands are thorny acacias and shrubs like the creosote bush. Tufted grasses sometimes grow in the shelter of such shrubs and thorns. Only a few deserts – in North and South America, and the Namib of southern Africa – support the "classic" desert flora of stem-succulents, which are cactuses and cactuslike plants. In herbaceous deserts grow smaller perennial, nonwoody herbs whose tops die back each year, but whose root systems can survive. Salt deserts support relatively sparse vegetation, though some sagebrushes and other shrubs can grow in areas of reduced salinity. Sand dunes are home only to plants with wide-spreading roots that can find an anchor and moisture in the loose grains.

In spring mature specimens of the stately saguaro cactus (*Carnegiea gigantea*) produce striking white flowers on their upper stems. Later sweet, red edible fruits appear, which are gathered by local Native Americans as food and to make an alcoholic drink.

### North American desert plants

Thanks to Western movies, classic desert plants such as cacti and agaves, which are common in the Mojave, Sonoran, and other deserts of southwest North America, are familiar to people who have never been to the region. Some of the cacti are capable of reaching tremendous proportions, especially the saguaro (*Carnegiea gigantea*) and the elephant cactus, or *cardón* (*Pachycereus pringlei*), both of which can exceed 15 meters (50 ft.) in height and weigh over 10 tonnes (9.8 t.). In desert environments they fulfil the same ecological role as trees elsewhere, providing perching sites and nesting holes for birds such as gila woodpeckers and elf owls (*see illus.*, p. 104). Their woody ribs are also used by people as firewood and building materials.

Cacti flower briefly but often to brilliant effect. The blooms are borne almost directly on the stem, on the specialized structures called areoles. Once mature, a cactus will flower annually, unless drought is exceptionally prolonged. The blooms are usually single and showy, varying from purples and reds to yellows, with petals and sepals united at

**Figure 3 Cacti come in a wide variety of shapes and sizes. The giant saguaro of southwestern North America – far right (not to scale) – is exceptional; most cacti are small and low-growing. Many are spherical in shape, which maximizes their volume but minimizes surface area, making them excellent water containers.**

their bases to form a tubular shape. Some, like the strawberry, or hedgehog, cactus (*Echinocereus enneacanthus*), and various opuntias, or prickly pears (*see panel*, p. 87), bear tasty fruits, which are sought after by animals and people alike.

Perennials of the North American deserts include the sagebrush (*Artemisia* species) – aromatic, low-growing shrubs of the daisy family and the state flower of Nevada – creosote bush, and brittlebush, all of which have small leaves to reduce transpiration and water loss. A characteristic sight of these deserts is the Joshua tree (*Yucca brevifolia*), a member of the yucca genus whose irregular branches point skyward with stiff, bayonet-shaped leaves and fragrant, showy white flowers borne on tall stalks.

The arid Southwest is also home to one of the world's most distinguished plants, the bristlecone pine (*Pinus aristata* or *longaeva*). This extraordinary tree grows between 3 and 15 meters (10–50 ft.) in areas where annual rainfall averages 300 millimeters (12 in.), including dry mountain slopes, notably in California's White Mountains. Above the twisted, distorted, often diagonal trunk are small branches densely packed with the typical "evergreen" leaves, or pine needles – except that the bristlecone keeps its needles for up to thirty years, compared to the two to three years of other pines. The needles are 4 to 6 centimeters (1.5–2.5 in.) long and are in groups of five. The woody cones are 5 to 10 centimeters (2–4 in.) long, and, as the name suggests, each scale is tipped with a long, thin,

**Bristlecone pines *(Pinus aristata* or *longaeva*) thrive in the most brutal conditions, exposed to bitterly cold winds and severe drought. Younger specimens of the bristlecone pine have more foliage; the very oldest are little more than blasted, twisted stumps.**

The massive stem of the elephant's foot yam, a native of arid regions of southern Africa, is a vital reservoir of moisture, not only for the plant itself but also for wildlife.

sharp, spiny bristle. The truly extraordinary feature of this tree, however, is its lifespan. Tree ring studies by dendrochronologists show that some specimens are almost 5,000 years old, vying for the title of oldest organisms on earth. This great age makes them useful to scientists calibrating carbon-dating techniques. They grow so slowly that they may take 3,000 years to reach full height, which in smaller individuals gives a growth rate of about one meter (39 in.) per millennium, or one millimeter per year.

## YUCCAS AND YUCCA MOTHS

There are about 40 species of yucca, a genus that belongs to the agave family. Many species, such as Spanish dagger (*Yucca gloriosa*) and Adam's needle (*Yucca filamentosa*), are today cultivated in well-drained parks and gardens throughout the world. In origin, however, the yucca is a native of arid areas of southwestern North America. Typically, a yucca has a rosette of extremely hard, swordlike leaves that sprout almost directly from the ground, together with a central stem, 3 meters (10 ft.) or taller, that bears night-scented creamy-white or violet flowers. The yucca's nocturnal scent is especially important because this attracts the small moths known as yucca moths (*Tegeticula* species, such as *T. yuccasella* and *T. pronuba*) that play a crucial role in the plant's pollination. A female yucca moth visits the flowers to gather pollen from the male stamens as food. If a particular flower is at a suitable stage, she bores a hole in its ovary at the base and lays an egg there. Then she climbs the flower's female part, the stigma, and deposits some male pollen, thereby pollinating the yucca. In return, the moth caterpillar is able to eat part of the yucca's seed-containing fruits. This extremely close relationship is another example of the mutually beneficial partnership between different species known as symbiosis.

### South American desert plants

Even the world's driest desert, the Atacama, receives occasional rain, and in this region botanical ephemerality is stretched to its limit, as poppies, speedwells, and other low herbs spring up and flower within days of any shower. In the Sechura and Patagonian deserts giant thistles, giant groundsels, and tough grasses colonize the arid, windy landscape. One of the most familiar is the pampas grass (*Cortaderia selloana*), which has been spread and bred around the world by horticulturalists. Its perennial, evergreen leaves are like long, slim swords, whose serrated edges are an excellent deterrent against grazing herbivores. Its tall, silky plumes of creamy spikelets appear in late summer.

Cacti are also numerous in South America, with what is probably the southernmost species of the group, the miniature *Maihuenia patagonica*, growing on the rocky plains of Patagonia. Many plants of the region, such as the three-lobed azorella (*Azorella trifurcata*) and llareta (*Bolax gummifera*), are dwarfs or adopt a "cushion" habit – that is, they are low-lying, with densely packed leaves and twigs – protecting themselves from the ceaseless, dry wind that blows down from the Andes. Their hairy, almost felted leaves, tightly packed in rosettes, also create a microclimate that prevents water loss.

### African desert plants

Few cacti occur naturally in the Old World, but in Africa the place of the taller, woody species is taken by euphorbias (genus *Euphorbia*) including spurges, which provide shade, bird perches, and tree holes for creatures to hide or nest in. Some species grow much lower, however. Clumps of the column-shaped, six-ribbed, spiky hedgehog euphorbia (*Euphorbia echinus*) reach only about 1 meter (39 in.) tall but perhaps 3 meters (10 ft.) across, so that they resemble a giant green-spined hedgehog crouching on the ground. Acacias (genus *Acacia*), or thorn trees, are widespread even in the continent's driest regions, including the small hashab (*Acacia senegal*), which is a source of gum arabic, used in the manufacture of adhesives. The blackthorn acacia (*Acacia mellifera*) tends to grow in the harder clay soils between the sand dunes, where moisture is not too far below the surface.

In various regions of Africa, including the Kalahari Desert, the camelthorn acacia (*Acacia erioloba*) sends roots some 30 meters (100 ft.) into the soil to gather precious water. A single tree may supports colonies of acacia rats, social weaver birds, and bush babies. The camelthorn is especially useful to animal life since it produces its yellow pom-pom flowers just before the rains rather than just after, when other vegetation is at its brownest and most withered. The blooms ripen to thick, half-moon-shaped pods, each containing hundreds of seeds, which are food for many animals, from mice to giraffe and gemsbok. The seeds pass through the digestive tracts of larger herbivores unharmed, and ready-wrapped in their own lump of fertilizer – a common way of spreading the species to new areas.

The shepherd's tree (*Boscia senegalensis* and others) has sweet-tasting leaves and flowers and summertime berries that are an excellent source of food for many creatures, from pied babblers to black-back jackals. In older trees the wizened trunks act as reservoirs for pools of rainwater.

Below the ground many plants build up stores of moisture and nutrients – especially energy in the form of starch – for tough times ahead. The yams (genus *Dioscorea*) of the climbing vine family have massive tuberous roots and in wetter regions thrive as crops, but some are adapted to aridity. The elephant's foot yam (*Dioscorea elephantipes* or *D. sylvatica*; *see illus.*, p. 83) grows even larger than its name suggests, weighing as much as 50 kilograms (128 lb.) when conditions are favorable. Ground-squirrels, mole-rats and similar rodent burrowers nibble regularly at these huge storehouses without destroying them, thus ensuring a future food supply.

The protein-rich morama bulb, or gemsbok bean (*Tylosema esculenta*), grows to a vast size, perhaps 250 kilograms (40 st.), and can hold as much as 200 liters (350 pt.) of water. The window plant (*Fenestraria rhopalophylla*) of the Namib Desert attempts to elude herbivores by having its small columns of fleshy leaves buried upright in the soil. Only the flat, translucent tops show, like vegetable "windows," above the surface, allowing light down into the leaves below for photosynthesis.

**"Living stones," such as this variety of *Lithops herrei* found in the Namib Desert, blend into the dry desert surface as a way of escaping the attentions of hungry herbivores.**

Grasses thrive in many areas where at least some rain falls. The Kalahari plume grass (*Stipagrostia kalahariensis*) is similar to the South American pampas grass, with feathery white heads releasing small windblown seeds. In the Sahara grows desert, or camel, grass (*Cornulaca monacantha*), a valuable fodder for camels, while toward the south are grasses more characteristic of the

## THE WELWITSCHIA: TWO LEAVES FOR 2,000 YEARS

The arid coasts of Namibia are home to one of the world's truly bizarre life forms. This is the welwitschia (*Welwitschia mirabilis*; *right*), named after its discoverer for science, German botanist Dr Friedrich Welwitsch, in 1860. It is a cone-bearer, or gymnosperm, and has a large stem mostly below ground, with a huge tap root that penetrates deeper still. The top of the stem usually projects only 10 to 15 centimeters (4–6 in.) above the surface, yet it may be more than 1 meter (3 ft.) in diameter, and it bears just two leaves. These grow from their bases to a length of several meters, and in the desert sand their tips become divided, threadbare, and shredded. The two leaves persist for the plant's entire life, which averages some 500 years and which in certain specimens exceeds 2,000 years. The welwitschia draws up water from deep underground and also soaks in the sea mists and heavy dews common along the Namib's shores.

**A pistachio tree has roots that stretch out from the plant up to 30 meters (100 ft.), while the nuts are adapted to attract feeders who will scatter the seeds widely, thus improving the chances of germination.**

scrubland of the Sahel, such as cram-cram (*Cenchrus biflorus*). Toward the north and west of the Sahara, approaching the Mediterranean and Atlantic coasts, are alfa, or esparto, grass (*Stipa tenacissima*) and diss grass (*Ampelodesmos mauritanicus*). These grasses grow rapidly in the rainy season, using their established root systems and sending out temporary roots within hours of the soil moistening, to absorb as much water as possible. Their roots also produce sticky secretions that bind together the surrounding sand and soil particles, helping to stabilize the ground against the swirling winds.

Some of the continent's most unusual plants do not look like plants at all, or even as if they are alive. They are the "living stones," such as *Lithops pseudotruncatella*, *L. lesleie*, and species of the *Gibbaeum* and *Pleiospilos* genera. The *Lithops* species have two domed leaves that are rounded, mottled, and muted green-gray or green-brown to resemble the smooth, windblasted pebbles among which they grow – a perfect camouflage that helps protect them from the attentions of desert herbivores. The *Lithops* flowers, on the other hand, are showy whites, yellows, oranges, and reds, fiery beacons to passing insects that carry out the plants' pollination. Since such blooms lose water rapidly into the dry air and their bright hues draw the attention not only of useful pollinating insects but also of hungry herbivores, they generally last only for a day or two.

Ephemerals, or "desert weeds," abound in Africa after rain, including the ubiquitous poppies (genus *Papaver* and other genera), goosefoots (genus *Chenopodium* and other genera), and various types of trumpet-flowered bindweeds or convolvulus, which can complete their life cycles in as little as three weeks. Tamarisks (genus *Tamarix*), baobabs, quiver trees, and other trees and bushes grow in many parts of the Sahara, offering shade to creatures and travelers.

## Asian desert plants

The Kara-Kum Desert of Turkmenistan and Kazakhstan has both patches of loose sand blown into dunes and, toward the west, areas of dark, salty clay soil – the desert's name itself means "Black Sands." Some species have adapted to both habitats. For example, the white variety of the saxaul tree (genus *Arthrophytum*) has tiny leaves suited to sandy soil, while the black variety grows on saltier soils and has no leaves for most of the year: the tree's greenish-white twigs and branches carry out photosynthesis instead. After the flowers develop fruits, and their seeds have been dispersed for the year, whole branches may fall off the tree to reduce evaporation.

The roots of the sand acacia – one of many psammophilic, or sand-loving, plants, able to withstand surface temperatures that soar to 65°C (150°F) in summer – dive into the subsoil to depths of 15 meters (50 ft.). Pistachio trees (*Pistachia vera*; *see illus.*, p. 85), 5 to 7 meters (16–22 ft.) tall and many centuries old, are another local drought-adapted species. Their lateral roots spread outward more than 30 meters (100 ft.), and the green-golden kernels, two-thirds fat and almost one-quarter albumen, are eaten by many lizards, rodents, birds, and other creatures.

Each year giant fennels (genus *Ferula*) shoot up 2 meters (6–7 ft.) tall, with thick, spearlike stems, bushy branch tops, and feathery leaves; as the summer drought deepens the plant withers back to a few centimeters high. Only once during its lifetime of six to nine years does the giant fennel bloom. It produces green-yellow flowers, whose winged seeds twirl up to 20 meters (65 ft.) from the parent plant, which is already dying after its lone reproductive effort.

Smaller flowers and herbs of the Kara-Kum include the almost obligatory carpets of red poppies after a spring shower. The region is also the natural home of the tulip (genus *Tulipa*), more than a hundred species of which – in a profusion of colors and patterns – are found in arid lands across Asia. Their bulbs store moisture, energy, and nutrients for fast growth in spring so that they are able to flower and seed before the onset of the long, dry summer. The flower's name is derived from Turkish word for "turban" – *tülbend* – which the bloom's shape resembles. Tulips began to reach Europe only after the visits of Flemish diplomats to the Ottoman court in the 16th century. In slightly damper areas hyacinths (genus *Hyacinthus*), pulsatillas (genus *Pulsatilla*), and blue varieties of iris (genus *Iris*), gentian (genera *Gentiana* and *Gentianella*), and larkspur (genus *Consolida*) seem to reflect the empty azure sky, interspersed with clumps of the rattling dry stems of feather grass (genus *Stipa*).

Farther east, the Gobi Desert is also home to tough grassy clumps, especially its namesake Gobi feather grass (*Stipa gobica*), as well as a type of wild rhubarb (*Rheum nanum*) and milkweed (*Astragalus pavlovii*). The desert's perennials include autumn-flowering crocuses (*Colchicum laetum*), the aptly named Chinese lantern (*Physalis alkekengi*), with its red, lanternlike blossoms, and the foxtail lily (*Eremurus spectabilis*). In the saltier soil grow Asian tamarisks, such as *Tamarix ramosissima*, various thorn trees, sagebrushes, and short-leaved goose-foot (*Anabasis brevifolium*).

However, large parts of the Gobi have been swept bare of soil by the incessant winds, and they are some of the most deserted places on earth, with very little or no vegetation able to survive. Across the whole of the vast, harsh Gobi Desert, there are only about 600 species of flowering plants. By contrast, the nation of Israel, itself parched in many areas such as the Negev Desert, is one-sixtieth the size of the Gobi yet home to four times as many native plant species.

### Australian desert plants

Vast tracts of the arid Australian outback are dominated by mulga acacias and mallee gum trees, or eucalypts. One of the stranger sights in dry areas of the Northern Territory are Livingstone palm trees growing in dry soil – relics from a time when the climate of the area was much wetter. Various types of low, wide-crowned, flat-topped, dry-adapted desert and scrub wattles (types of acacia) also cover large areas. Some are known as ironwoods or axe-breakers due to the great hardness of their wood. The common ironwood (*Acacia estriophiolata*) begins as a prickly, squat shrub but slowly develops into a tree some 10 meters (33 ft.) tall, with drooping branches and narrow green-gray leaves, giving an uncanny resemblance to the weeping willow of lush riverbanks. Similar in name, ironbarks are not acacias but gums (genus *Eucalyptus*). Most types have blue-grey leaves and hard bark, which is dark, rough and deeply furrowed, as in the mugga (*Eucalyptus sideroxylon*).

Wildflowers carpet the dry outback after rain, including many kinds of daisies and wild cabbages. Flannel flowers (*Acinotus leucocephalus*, *A. helianthii*, and others) have daisylike heads with a golden center surrounded by white petals, together

## LIFE AFTER FIRE

In the deep desert fire poses little risk. Conditions are ideal for combustion – hot and dry – but there is almost nothing to burn. Any plants present are usually too widely spaced to catch alight one from another and spread the flames. In arid grassland, scrub and bush, however, fires are so common as to be a natural, and sometimes essential, part of the ecological system.

In Africa evolution has designed various types of terminalia trees (genus *Terminalia*) to exist mainly in the form of perennial roots. A slender stem grows from the root at ground level each year and sprouts leaves, but is burned away by the regular dry-season bushfires, usually started by lightning strikes from dry thunderstorms. However, the root, safe underground, continues to enlarge and become more woody. Sometimes, however, there is no annual fire, and the main stem grows at an amazing rate, sprouting leaves at a height of 2 to 3 meters (6–10 ft.). The following year a sapling tree is finally established, its thick bark able to resist flames and its tender leaves out of major harm's way.

In Australia the "blackboy" or grass-tree (genus *Xanthorrhoea*) of the outback flowers only when stimulated by the great heat of a blaze. This curious grass plant has a short "trunk" composed of the fibrous bases of its leaves, while the long, thin blades themselves extend above and to the sides in a clump, like the spines of a green hedgehog. After their scorching, the blackboys each produce a tall, columnlike flowering spike of pale yellow, to exchange windblown pollen and produce the seeds of the next generation.

# THE CURSE OF THE PRICKLY PEAR

The 200-plus species of opuntia cacti (genus *Opuntia*) have their homelands in arid parts of the Americas. They have been bred for their hardiness, their beautiful flowers in yellows, reds, and purples, their curious stem shapes, and their prickly, spiky hedging qualities, and some species have become naturalized in other dry areas such as parts of Europe. Many species are known as prickly pears, the commercially grown Indian fig, or Indian tuna (*Opuntia ficus-indica*), yielding pear-shaped fruits with sweet red or purple flesh.

The most spectacular opuntia "export" has been the flat-stemmed prickly pear (*Opuntia inermis*; *below*). This was introduced into southeast Australia in the 1830s, chiefly as a hedging plant to keep unwanted herbivores out of parks and gardens, to provide protective thickets for breeding game birds, and as food for cochineal insects. Since, however, the cactus found no natural enemies or eaters in its adopted land, it soon out-competed the local vegetation and rapidly spread. By the 1880s it had broken through the Great Dividing Range, which runs along the eastern edge of the continent, and was heading westward into the outback. By 1925 some 25 million hectares (62 million acres) of grazing and other land in Queensland and New South Wales were infested by this prickly weed. Salvation arrived in 1926 in the form of a biological control agent – the light-brown cactoblastis moth (*Cactoblastis cactorum*) from Argentina. Its orange-red, black-spotted caterpillars munch their way through the cactus's flesh until the hollowed-out plant collapses. Over the following years moth and cactus battled it out and the curse of the prickly pear was gradually lifted.

with deeply indented, almost furry leaves to keep out the dryness and heat. They thrive in sandy soils under the shade of gum trees. In the northwest the hairy pink and mauve blooms of mulla-mulla (*Ptilolotus exultans*) make a bright sight over the mulga scrub, flowering and fruiting within a few weeks of the winter rains.

A characteristic grass of arid Australia is spinifex (genus *Triodia*), also known as porcupine or tussock grass and not to be confused with "true" spinifex (genus *Spinifex*), which are creeping coastal sand-dune plants. Its spiky leaves are long and dry and form tall, dense clumps many meters across. A large clump can harbor dozens of animal species – from wolf spiders and desert centipedes to reptiles such as the bearded dragon and jeweled gecko, birds such as the orange chat, and small marsupials including the hare-wallaby. Some creatures, such as the spinifex snake-lizard and spinifex hopping-mouse, are rarely found elsewhere.

# CREATURES OF THE DESERT

*In the heat of the day a desert can seem devoid of life, and yet this harsh environment in fact supports a diversity of fauna. Insects and reptiles flourish in shady crevices, while small mammals take refuge underground. It is at night, though, that the desert really comes alive.*

The animal life of the desert depends on its plants, which are essential stores of nutrition and, more importantly, of water. Greenery – or as is more often the case in desert environments, brownery – is essential for maintaining a constant supply of water. Some of the world's greatest diversity and sheer numbers of creatures occur in the moistest environments of the rain forests. Humans, too, tend to live in moderate climate zones with plentiful supplies of water; in normal conditions the human body needs to take in about 2.3 liters (4 pt.) of water a day, equivalent to some 4 to 5 percent of total body weight.

Desert environments cannot support such a luxurious fluid intake, and animals in arid regions are adapted to reduce their water requirements to as little as 1 percent of total body weight. Such adaptations are generally one of three major kinds. Behavioral adaptations include such strategies as nocturnal foraging, which enables many small desert animals to reduce water loss by avoiding the dehydrating effects of hot sun. Anatomical adaptations might include a scaly covering to reduce moisture evaporating from the body, while a common physiological adaptation is the production of small volumes of very concentrated urine and dry, cakelike feces.

**The long legs of a darkling beetle (genus *Tenebriondae*) keep its body well clear of the burning desert surface. Animals of many kinds have evolved adaptations that enable them to survive the desert's perilous conditions.**

As in any habitat desert life organizes into systems of feeding links called food chains, which interconnect as food webs. At the base of every chain are plants – called primary producers by ecologists – which provide nutrients, minerals, and energy for the next link or level, the herbivorous animals. In the desert environment these vary from ants, beetles, and bugs to small rodents and seed-eating birds and to larger mammals such as gazelles and asses. Herbivores in their turn become meals for the carnivores – the third link in the food chain. Smaller carnivores such as lizards, spiders, and insectivorous birds then become prey for larger predators, known as the top carnivores, such as cheetahs and hawks. These are not, however, the final link. All animals, even top carnivores, eventually die and provide nourishment for the numerous desert scavengers in what are called the detritivore links of the food chains. In this way nutrients and minerals are passed on, either directly to other animals or by rotting back into the ground as a natural compost for fresh plant growth. Even in the desert nature endlessly recycles the raw materials of life – from the soil to plants to animals and back into the soil. Nevertheless the harsh, sparse nature of the desert means that food chains and webs are severely restricted, both in terms of length and the number of interconnections. There are simply fewer species of living things; there is, as scientists say, a low biodiversity.

## Strategies for Survival

Two groups of animals – arthropods and reptiles – are particularly well adapted to the desert environment and play a dominant role in its food chains. In the desert environment arthropods, whose hard body casings help to conserve precious body water, are represented mainly by insects, such as grasshoppers, flies, beetles, bugs, ants, and termites, and arachnids, such as spiders, scorpions, and mites. The principal types of reptile found in deserts are lizards and snakes, whose scaly covering, like the arthropods' carapace, restricts water loss, unlike the thinner, more exposed skin of amphibians, birds, and mammals, all of which flourish less well in arid conditions.

Bodily water conservation is only one reason why arthropods and reptiles are so successful in arid habitats, however. Body temperature, too, plays a crucial role. Deserts are places not only of extreme heat but also of extreme cold, with nocturnal temperatures that can often dip toward freezing. Both temperature extremes present life-threatening challenges to the maintenance of steady body temperature, in particular for warm-blooded vertebrates such as mammals and birds. To survive the desert night, mammals, for example, must find sufficient and continuing food supplies to act as fuel to maintain a healthy body temperature – from 35°C to 43°C (95–110°F) according to the species – and food, like water, is in short supply in the desert. Moreover, during scorching daylight hours mammals' cooling systems are triggered to prevent overheating, involving the loss of precious body fluids as the animal sweats or pants.

Unlike mammals, arthropods and reptiles are not warm-blooded but poikilothermic; that is, they have a variable body temperature that is generally slightly higher than the ambient temperature. For example, in the desert sun their body temperature may exceed the 40°C (104°F) or so of a bird or mammal. Many poikilothermic animals can regulate their body temperatures by behavioral means, seeking out shade and cooling breezes when it is too hot and basking on a heat-retaining rock when it is too cool. The great advantage of poikilotherms is that they require far less energy and, hence, food compared to warm-blooded animals. Some experts estimate that poikilothermic animals like reptiles and arthropods need about five or six times less food than mammals relative to their body weight.

### Making moves

In addition to extreme temperatures and sustained drought, desert-dwelling animals have also had to adapt their movement to the peculiarities of the desert terrain. The classic terrain of the desert is sand, and this presents hindrances to free movement familiar to anyone who has taken a walk on coastal dunes. Animals of sandy deserts experience the same problem and have developed an array of methods, including anatomic and behavioral adaptations, for moving more easily over this shifting and blazing-hot substrate.

- Hopping or leaping. Powerful hind legs and large, long rear feet can leap over the substrate effectively. The surface area of the rear foot spreads the animal's weight over a long strip oriented in the direction of travel, causing a reduction in the rearward slippage of soil particles or sand grains on takeoff. This method of movement is used by various mammals, from jerboas to kangaroos.
- Sandshoes. A large surface area on feet helps to prevent the body from sinking into soft sand, much as snowshoes do in soft snow. The broad feet of the camel perform this function, while many types of desert lizard, such as the whiptail and collared lizard, have very long toes, often with fringes or large scales that similarly extend body-to-sand contact.
- Shuffling. While walking, diurnal desert lizards often shuffle their feet at each step, pushing aside the very hottest surface sand grains and

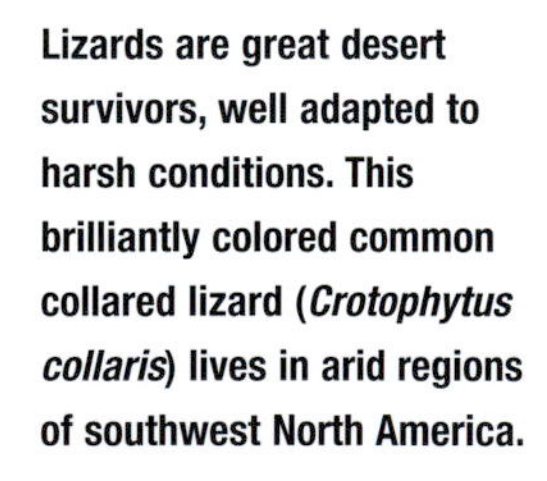

**Lizards are great desert survivors, well adapted to harsh conditions. This brilliantly colored common collared lizard (*Crotophytus collaris*) lives in arid regions of southwest North America.**

remarkable emergency escape mechanism. By curling its legs together under its body to form a cartwheel or ball shape, it is able to roll at speed down a dune or hillside.

## Behavioral strategies

So far we have primarily seen how physiological and anatomical adaptations enable animals – especially reptiles and insects – to survive extreme desert conditions. Behavioral strategies, however, are also vital in the battle against heat and drought, particularly for mammals, which are anatomically less well adapted to the desert environment.

Many species escape the heat of the day by adopting nocturnal habits, hiding in burrows and tunnels in the sandy soil or in caves and crevices among desert rocks – all places that have more equable microclimates. Out in the full glare of the sun, the air temperature may be 45°C (113°F) or higher and the ground temperature an almost boiling 90°C (194°F), whereas in a cool burrow just 10 centimeters (4 in.) below the desert surface, the temperature may be reduced to 30°C (86°F). This burrowing strategy is adopted only by smaller animals. Any animal larger than, say, a rabbit, would have to dig its burrow deeper, where the soil is usually harder and more compact, or where loose earth tends to collapse. Such a burrow would, in any case, be likely to admit predators.

A huge variety of small desert mammals use burrows, including rat- and mouselike rodents such

**A desert sidewinding viper (*Bitis peringueyi*) sidewinds across sands in the Namib Desert. This distinctive method of movement both minimizes body contact with the scorching desert surface and facilitates the swift pursuit of prey.**

allowing their feet to rest briefly on the slightly cooler grains beneath.

- "Swimming." Numerous species of snakes and lizards are able to "swim" through very soft sand or loose soil, making sideways undulating movements of the body to tunnel just below the surface. The sandfish lizard (*Scincus scincus*) of Middle Eastern deserts is expert at this type of propulsion.
- Sidewinding. The sidewinder rattlesnake (*Crotalus cerastes*) of North American desert regions is named for its distinctive method of locomotion, which leaves in its wake a striking trail of elongated S- or J-shaped marks across the sand. The sidewinder moves by throwing its body into a series of sideways waves, lifting its head from the substrate, throwing it diagonally forward, and then setting it down again several centimeters farther on in the direction of travel. This elaborate method of propulsion allows the snake to push rearward with its whole body length against the sand for extra grip or purchase and with each loop to lift every part of the body clear of the hot sand, giving precious cooling-down time.
- Rolling. This method of movement is unique to the solifuge, or sun spider (*Galeodes arabs*, for example; *see also* p. 98). The solifuge has a

## METABOLIC WATER

One physiological survival strategy is the exploitation of metabolic water. Most animals lose more water than they take in. The extra water is made by the biochemical pathways during digestion.

In the desert metabolic water can be a valuable addition to external supplies. Metabolic water is a natural product of the chemical breakdown of food in the body. Animals gain energy from the sugars they derive from digested food, especially glucose or blood sugar ($C_6H_{12}O_6$), which are broken down using the oxygen (O) absorbed from inhaled air. The products of this process are not only energy, but also carbon dioxide ($CO_2$), which is exhaled, and water ($H_2O$).

Metabolic water is also produced when breaking down long-term stores of food, especially fats. This is why so many desert creatures, from gila monsters to camels, use fat stores, which they use as reserves of both energy and water.

as kangaroo rats, gerbils, jirds, pocket mice, and pack rats. Some of the desert-dwelling ground squirrels, for example, excavate complex burrows up to 20 meters (66 ft.) long, connected by cross-links into a network with eight to ten entrances and several chambers down to 2 meters (78 in.) below the ground. Similarly the crassus jird of the Sahara excavates a system with a combined tunnel length of some 40 meters (130 ft.), more than 15 entrances, and several chambers serving as nests, nurseries, and food storerooms up to 1.5 meters (5 ft.) below the surface.

Burrow dwellings have other advantages besides their relative coolness. Tunnels function as relatively safe foraging areas, allowing desert animals to find food with a diminished danger of themselves becoming food. Ground squirrels, for example, feed on roots, corms, bulbs, tubers, and other underground plant parts, while desert moles feast on worms, beetles, termites, and other small creatures that fall through the walls of their subterranean galleries. A second advantage of the burrow is the increased humidity found only a few centimeters beneath the dry desert surface, causing a reduced loss of body moisture.

Desert reptiles, such as skinks and similar lizards, and invertebrates, such as scorpions, also make extensive use of caves, cracks, and burrows to hide from the drying sun and wind as well as from predators. They tend, however, to occupy ready-made natural hollows or old and abandoned rodent tunnels rather than excavate their own.

## Staying cool

Some desert-dwelling termites, or "white ants" as they are sometimes misleadingly known – their closest relation is in fact the cockroach – have an alternative solution to the problem of staying cool. These pale, moist creatures are particularly vulnerable to high temperatures, drying out and dying within minutes in hot sunlight. Most termite species live in colonies and excavate underground nests where they live in cool, dark dampness. Some desert species, however, also gather and pile desert soils to build towering structures of earth – termite mounds – that can sometimes exceed 5 meters (16 ft.) in height (*see* Fig. 1 *opposite*). Termite mounds rapidly become baked hard in the heat and are a conspicuous feature in many arid lands, especially in the deserts of southern Africa and Australia.

The termite mound or tower does not contain the termites' nest. The actual dwelling area is below ground level as usual, directly under the mound, with feeding galleries radiating into the surrounding soil, where the termites forage for fragments of plants, fungi, and other foods. Instead, the mound serves as protection, especially against termite-eaters such as the giant anteater and against flash

**The sand cat (*Felis margarita*) inhabits arid areas of the northern Sahara, the Arabian Peninsula, Pakistan, and, possibly, Iran. Sand cats hunt by night and use their large ears and well-developed hearing to locate prey. About the size of the domestic cat, they feed on birds, small mammals, reptiles, and locusts.**

floods from rare cloudbursts. The mound also serves as a natural air-conditioning plant. A network of holes, ducts, and chimneys allows air to circulate freely, drawing heat away from the nest during the day – though without taking too much valuable moisture – while preventing the nest cooling too quickly at night.

Termite mounds are a useful feature of the desert landscape and provide in miniature a paradigm of the interdependence of life-forms in the desert environment, and indeed in any biome. Birds use them as perching and vantage points. Cheetahs or kangaroos lie in their shade. Snakes, armadillos, jackals, or other den-users may inhabit parts of the mound that the termites have deserted. In Australia mulga parrots (*Psephotus varius*) often peck a nesting hole in a termite mound rather than a tree, while sand goannas (*Varanus gouldii*) tear open part of the mound to lay their eggs inside. The termites subsequently repair the wall, providing a haven for the lizard's eggs to develop within.

Some of the most intriguing termite mounds are the slab-sided, sharp-topped, squared-off structures found in the arid regions of northern Australia. Up to 4 meters (13 ft.) tall, they look like the tips of giant chisels sticking up from the hard, red earth. Their builders are known as magnetic, or compass, termites (*Amitermes meridionalis*) because the broad sides of the mounds are always oriented north–south. This remarkable phenomenon is due, however, not to any magnetic-detection skills on the part of the termites but to the fact that they build the mound in such a way as to maintain an even temperature both within and beneath the structure for as long as possible each day. At dawn and dusk the sun shines on a broad side of the mound, keeping it warm for as long as possible, but at midday the sun blazes down on the small surface areas of the sharp edges and top, preventing overheating.

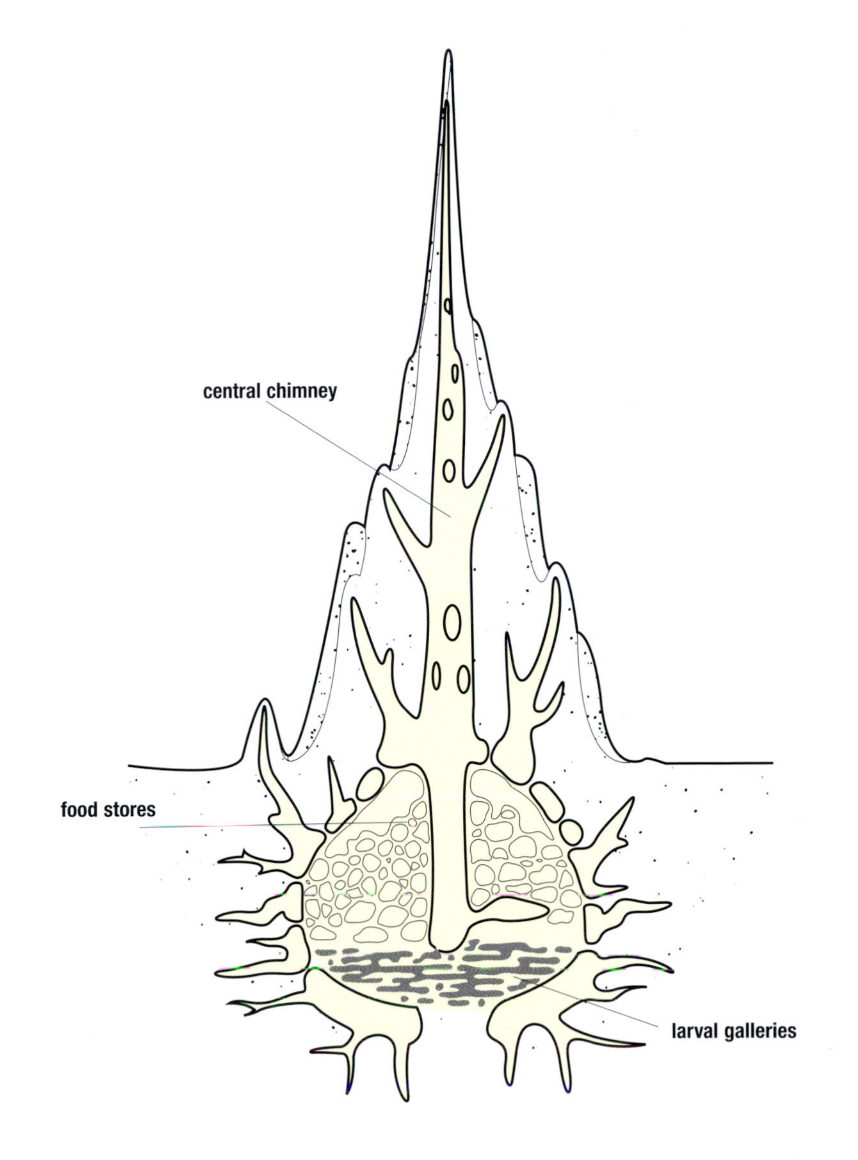


**Figure 1 Termite mounds are often dominated by a large central chimney that acts as a hot-air vent, keeping the nest cool in the desert heat.**

## Estivation and migration

In cooler, more temperate lands many animals become inactive and hibernate through the harsh season of winter. In deserts a similar "inaction strategy" – known as estivation – allows creatures to survive the worst of the heat and drought. Snails, for example, use their mucus to stick their shells to a hard substrate such as the underside of a shady rock. The mucus dries, creating a waterproof seal round the shell and keeping the snail moist inside for months. Some desert amphibians, such as spadefoot toads (genus *Scaphiopus*) and water-holding frogs (*Cyclorana platycephalus*; *see* pp. 102–103), burrow deep into soil to estivate through the dry season.

Bats may not seem to be typical desert animals, but of the more than 900 species, some two dozen can survive in arid, treeless landscapes. The great mouse-tailed bat (*Rhinopoma microphyllum*) of the Middle East and East Asia, for example, roosts in cracks and caves in rocky outcrops or in the eaves of ruined tombs and temples. It feeds on all kinds of insects and, as the dry season approaches, lays down extra layers of body fat before settling in its roost and entering an inactive or torpid state that is a combination of hibernation and estivation. It subsists like this for two to three months on its stored food, waking up at almost half as heavy as before its quasi-sleep.

The sparse and unpredictable nature of food availability in deserts has led to many animals evolving nomadic or migratory habits. In some less arid areas, regular migrations are tied in with dependable seasonal rainfall, but in true deserts, where such rains are lacking, migration is more haphazard. The zeren gazelles (*Prodorcas gutturosa*) of the Gobi trek to new pastures mainly in spring and fall and may cover hundreds of kilometers, but the direction they take depends entirely on the weather and where the pastures of feather grass and other plants are growing. In winter they may have to move on every two or three days.

## RARE SURVIVORS

Crustaceans are an important group of mainly sea animals that include barnacles, shrimps, prawns, crabs, and lobsters, as well as freshwater "pond fleas," or daphnia. The existence of such creatures in the desert may seem unlikely, but, remarkably, some species – including the shield shrimp (*Triops australiensis*; *below*) of Central Australia and fairy shrimps – have found a way to survive in this near-waterless environment. They may spend years as tiny, hard-cased eggs, scattered and mixed into the desert soil, until with the advent of rain, the eggs hatch within a few hours.

Once hatched, the shrimps grow at a remarkable rate, feeding on microscopic and tiny life-forms and thriving in the temporary rainwater pools and puddles. Within days they are almost fully grown and are producing thousands of their own eggs, as well as providing food for other desert water-lovers such as frogs and toads. The shrimps' whole life cycle may last less than two weeks before the pools dry and the eggs settle back into the sand. Brine shrimps (genus *Artemia* and others) are mostly fairy shrimps specialized to live in salty water, which occurs in desert salt pans as found in deserts from the Mojave to the Kalahari.

Perhaps equally startling inhabitants of the desert are the fish that live in rare and transient desert streams. These near-miraculous survivors are delicately adapted to their warm and mineral-rich habitat, and their existence is precarious indeed. In the arid southwest of North America several types of pupfish (genus *Cyprinodon*) live in such streams and cavern pools, including one of the world's smallest, least numerous, and most restricted fish species – the Devil's Hole pupfish (*Cyprinodon diabolis*) of Nevada – which is confined to a single underground pool.

### Desert locusts

Many other desert creatures follow the same irregular nomadic lifestyle in search of food and water – from birds such as emus and budgerigars in Australia to large mammals such as the asses and gazelles of the Sahara. One of the most notorious examples of these unpredictable migrants is the desert locust (mainly *Schistocerca gregaria* and *Locusta migratoria*, but also many others), whose arrival en masse from the deep desert inspires terror among the inhabitants of the tropics. A major swarm can number 10 billion of these insects, turning the sky black for hours and eating every shred of greenery, including precious crops. These locust irruptions, or population outbursts, are less common today than 50 or 100 years ago, and early-warning systems, along with detailed weather forecasts, spotter aircraft, and crop sprays, have helped to stem their devastation. Now and then, however, unexpected swarms can still inflict damage, bringing hunger to large areas. Worst affected are Africa and the Middle East, but many other regions, including the Americas and Australia, also experience periodic irruptions.

The young, or larval, forms of locusts, just hatched from eggs that their mother laid buried in the sand, are called nymphs. These locust nymphs, whose wings are present but too immature for flight, are known as "hoppers," and this is what they do. Hopping over the dry scrub and bush, like their close relatives the grasshoppers and crickets, they use their hard mouthparts to cut and grind up grasses and other plants. They are well camouflaged in blotchy greens and browns but still form staple food items for many desert insect-eaters, such as lizards, toads, and grasshopper-mice.

In most years, when the dry season persists as usual, vegetation is in short supply and puts a

**Desert locusts swarm the ground in Saharan Mauritania. Locusts can destroy crops over a wide area, threatening human populations with famine.**

natural ceiling on hopper numbers. A spell of extra rainfall, perhaps just a week longer than normal, means a boom in greenery. Along with many other herbivores, the locusts thrive, growing faster and completing their life cycle within two months. As the hoppers become more numerous, they also change appearance, swapping their discreet camouflage for bright stripes of yellow, black, and orange, entering what is known as their gregarious phase. Wearing their new, easily spotted colors, they gather into bands of thousands that march across the landscape, one or two kilometers a day, devouring every scrap of vegetation. They feed at dusk and dawn and rest during the heat of midday.

After passing through the fifth and final skin-moult, their wings are fully formed and, as adult locusts, they take to the skies in search of fresh pastures. As soon as the dryness returns, the remaining locusts, no longer stimulated by plentiful food and the presence of thousands of their kind, revert to their restrained solitary phase once more.

## A Desert Bestiary

The diversity of the world's deserts, in terms of climate and vegetation, means that each has evolved a different fauna. For example, the near-waterless Taklimakan supports very little wildlife while, by contrast, deserts such as those in the American Southwest, with their varied terrain and vegetation, have a comparatively rich fauna. The deserts of the island continent of Australia support some of the planet's most remarkable wildlife, the numerous species of desert-adapted marsupials.

### Insects of the desert

As in virtually every terrestrial habitat insects are the most common creatures in a typical desert. The planet's most brilliant survivors, they are able to adapt to the harshest of conditions. Their small size means they can access a wide variety of microclimates – such as those provided by cool and moist nooks and crannies – many of which reptiles, birds, and mammals are unable to exploit. The insect "skin," or cuticle, is waxy and waterproof and resists moisture passage and desiccation so effectively that some species, especially beetles, are

**A giant desert centipede (*Scolopendra heros*) attacks a tarantula using poisonous fangs that have been adapted from its foremost pair of legs. Many desert creatures have evolved extremely toxic venoms, which help ensure that rare prey does not get away easily.**

## THE ANT LION

The ant lion (genus *Myrmeleon*) is not an ant at all but, along with lacewings, a member of the group of insects known as the Neuroptera. The adult ant lion resembles a dragonfly and is a fierce aerial predator. The larva has fanglike jaws, which are among the largest, for its body size, of any creature. Desert species of ant lions live in loose soil or sand, excavating cone-shaped pits where they hide at the bottom just below the surface. An ant or similar victim stumbles and slides into the pit, unable to get a purchase on the loose grains. The ant lion larva seizes the prey, sometimes tossing aside soil particles so that the prey scrabbles but still slips downward toward the massive jaws. The ant lion drags its freshly caught victim under the surface, pierces its body, and sucks out its juices.

able to travel many meters in the midday desert sun without sustaining damage. Insects also perform a vital role as interstitial links in the many food chains that connect plants to small carnivores.

Ants, like termites, are especially widespread in deserts. Despite belonging to two different insect orders, respectively the Hymenoptera and Isoptera, both ants and termites have superficial resemblances in terms of body size, shape, and anatomy; in their social organization, with castes of workers, foragers, guards, courtiers, and the queen; and in nest-building for their colonies. Ants, however, are hardier than termites, able to cope with even hotter and drier conditions, and to wander farther from the nest on food-finding missions. Some ant species, for example, are known to venture into the surrounding hostile landscape for more than 50 meters (164 ft.).

Foraging hymenopterons gather all kinds of food scraps to take back to the colony and share out. Harvester ants (genus *Pogonomyrmex* and other genera), for example, specialize in collecting the tiny seeds of dryland grasses and similar plants, which they store in piles in underground, granarylike chambers. A single harvester ant worker can amass 20,000 tiny seeds in one week, which is about four per minute during daylight hours – a Herculean achievement.

Some of the strangest desert ants are the honey, or honeypot, species (genus *Myrmecocytus* and other genera), found mainly in the deserts of North America, Africa, and especially Australia. Specialized workers in the colony known as repletes have evolved to act as living larders for both moisture and energy. As the dry season approaches, their bodies swell with a treacly fluid, which they process from honeydew and nectar foods delivered by foragers. The repletes store their wares in the part of the gut called the stomach or crop, so that their abdomens swell like amber-colored pearls to some 8 to 10 millimeters (0.4 in.) across. Too distended to move, the repletes lie on the nest floor or hang from its roof.

When moisture is scarce, the repletes are literally tapped for food by other workers in the colony. The worker pats the replete with its antennae, and the replete exudes some of its syrupy store, which the worker then laps from its mouth or anus. A large nest may have several hundred of these living food stores, usually in one of the deepest chambers. As the dry season fades, the repletes give up the last of their sweet honeydew. In some species they then resume normal nest duties, while in others they simply perish and their bodies are removed unceremoniously from the nest.

**As the drought season approaches, some members of honeypot ant colonies begin to act as living water storers, their abdomens bloating to many times their body size.**

While ants and termites have adapted well to the rigors of the desert, it is the beetles (Coleoptera – the earth's most prolific order of insects) that are the supreme drought survivors, with many thousands of species eking out their inconspicuous lives in the desert, quietly feeding and breeding. The majority scurry across the surface soil in their search for small edible scraps of plant and animal matter. A few beetle species, however, have higher profiles due to their more predatory habits.

Among the most conspicuous is the thermophilic, or "heat-loving," black ground beetle – one of the few creatures out by day, walking over the scorching surface in the blazing desert sun. It preys on other insects, grubs, and small creatures, seizing and chopping them up with its outsize jaws. The beetle's long legs give a good turn of speed and hold its main body above the hot desert sand. Hairs on the legs and feet detect vibrations both in the air, which could be an approaching predator, and in the ground, which might be a hidden meal. If challenged, the black ground beetle contorts its body around to spray a burning, stinging fluid from its rear end at the enemy.

The jewel-bright scarab (*Scarabaeus sacer*) – a type of dung beetle – of the Saharan edge was revered by the ancient Egyptians for its glistening colors and held a key place in their cosmology – they thought of the sun as being rolled though the sky by a giant scarab. The sacredness of the Saharan scarab was perhaps an early recognition of the importance of all dung beetles in the ecology of the desert. The recycling of animal droppings is especially important in the dry desert soil, where dung on the surface quickly loses its moisture and becomes too hard-baked and desiccated for the normal processes of microbial decay.

Dung beetles (*Scapabaeus*, *Canthon*, *Copris*, and other genera) chew off balls of the droppings and bury them in chambers below the desert surface, laying an egg on each one. Once the beetle grub (larva) has hatched, it eats the dung, which has become soft and edible after absorbing moisture from the surrounding soil. The grub evacuates its own droppings underground, thereby taking the breakdown and return of minerals and nutrients into the ground one stage farther. The grubs also pupate underground and emerge as adult beetles at the surface to continue their life cycle, leaving an enriched soil below.

Dung beetles have evolved to deal with the particular fecal matter of their region. In Australia most of the local large mammal species, such as kangaroos, are drought-adapted and so produce very dry solid wastes. The Australian dung beetles, such as the genera *Blackburnium* or *Bolboceros*, thrive on this, and their plump, white, C-shaped larvae, called curl grubs, are common even in dry grassland and arid scrub. When cows and sheep were introduced to these marginal pastures in the 19th century, the native beetles shunned their moist droppings and other dung beetles had to be imported from the homelands of the newcomers to maintain the ecological cleanup process.

Scorpions (order Scorpiones), eight-legged arthropods of the arachnid group, are classic desert animals, enduring extremes of temperature that would kill most other creatures, and can enter a state of inactivity or torpor during which they need

**Compared with most mites, red velvet mites (genus *Angelothrombium*) are very large, about 6–8 millimeters long. They remain in the soil most of the year, and typically emerge from the ground only after spring rains to feast on prey that also emerges at this time.**

## A DRINK OF DEW

The coastal Namib Desert of southwest Africa is characterized by sea fogs (*see* pp. 32–33) that may drift several kilometers over the narrow strip of arid land; the fogs are particularly heavy at dawn and dusk on days when gentle onshore breezes are blowing. Numerous plants have adapted to exploit the dew that the fogs create. Likewise, the darkling beetle, a member of the tenebrionid beetle group, has developed a unique method of exploiting this source of moisture. As the ocean mist rolls in over the sand, the beetle races to the top of a dune and tilts its body upward in a "headstand" so that the underside of the abdomen faces into the damp, slow-moving air. Tiny droplets of water collect on its smooth, shiny underside and roll down to its head, where the beetle literally drinks in the seaborne moisture.

no food or water for many weeks. Desert-dwelling scorpions are usually a sandy color for camouflage, although most species hide by day in burrows or under stones, emerging only at night to hunt. Their four pairs of simple eyes provide only poor vision, and their main senses are touch and smell. The hairs on the legs detect air current and vibrations, including sounds; the tiny time delay between vibrations reaching the nearer leg compared to the farther one tells the scorpion from which direction the sounds come. There are also featherlike parts called pectines on the insect's underside, not found in any other animal group, and these are thought to pick up vibrations from the ground.

The Saharan androctonus (*Androctonus australis*) and Saharan buthus (*Buthus occitanus*) are among the most poisonous of the scorpion order, injecting venom from the sharp-tipped tail sting that can kill a human or a largish mammal. The sting of the centruroides, or sculptured scorpion (*Centruroides sculpturatus*), of North American deserts – yellow or yellow-brown with two dark back stripes – is also potentially fatal. A threatened scorpion warns a potential aggressor by flexing its tail, holding the sting at the ready and raising its outstretched pincers. It may also, but does not always, use its poison on its prey. The scorpion grabs the victim with its two large, pincerlike pedipalps and tears it apart with the smaller but more powerful jawlike chelicerae. Large scorpions tackle insects, spiders, centipedes, lizards, and even small mice and rats.

The adaptation of very potent venom is shared by many desert predators, including desert spiders, snakes, and wasps. In the desert environment, where prey is often scarce, powerful, fast-acting venoms not only improve the chances of capture, but help to avoid a long pursuit through the dry heat. Toxins are used for defense as well as attack, however, and a variety of small desert creatures, including whip-scorpions, blister beetles, millipedes, and mites, spray out distasteful, stinging, or burning secretions when in danger.

Velvet mites (family Thrombidiidae) – an arachnid about the size of housefly with a red, velvety covering – wander the desert sand eating any scraps of plant or animal matter they can find. Natural toxins in their bodies are so poisonous that they can kill small carnivores such as grasshopper-mice that disturb or attack them.

## THE FEARSOME SUN SPIDER

The sun, or wind, spider (*Galeodes arabs*, for example), also called the solifuge, is a powerful and voracious predator, found chiefly in desert and arid regions across Africa, Asia, and central and southern North America. Although not a true spider, it closely resembles one, with an outsized head and massive fangs. It lacks a poison bite and instead hunts by racing at speed over the substrate and jabbing its fangs deep into its prey, which might be a grasshopper (*as shown below*), a true spider, lizard, or even a small rodent, such as a mouse. The solifuge's head and body reach 5 centimeters (2 in.) in length, and with its spread legs it may be the size of an outstretched human hand. Six legs are specialized for fast running, with the slimmer front pair being used as feelers in the manner of an insect's antennae.

## DESERT TORTOISES

While the chief reptile groups represented in deserts are snakes and lizards, tortoises also occur. Their thick shells reduce water loss greatly as well as affording protection, and their lugubrious movement means that panting, which gets rid of water vapor from the lungs, occurs only exceptionally. In Africa the leopard tortoise (*Geochelone pardalis*), with its high-domed shell sporting blotches that resemble a leopard's spots, dwells in savannas and on desert fringes. If the ground is too hard, the female urinates on it so she can dig a hole for her eggs. The African pancake tortoise (*Malocochersus tornieri*) has an unusually flat shape, as its name suggests, and a highly flexible shell. Instead of withdrawing into its shell when threatened, it runs – quite fast – to a crack or crevice and inflates its body with air so that it becomes wedged in and immovable. The gopher tortoise (*Gopherus polyphemus*) of North America has wide, flattened, spadelike forefeet to dig a burrow up to 14 meters (45 ft.) long. It basks and chomps by day and hides by night in the chamber at the end of its tunnel, the way blocked by its hard shell. The species of desert tortoise shown right, eating the moist fruit of the prickly pear, is *Gopherus "Xerobates" agasszii*, a tortoise found in the Sonoran Desert.

### Reptiles of the desert

The reptilian body, with its scale-covered, thick, leathery skin and low energy requirements, is in a sense preadapted to desert conditions. Unlike mammals, too, it does not sweat and produces thick, pasty bodily wastes, which minimizes water loss. A typical desert reptile, such as the desert night lizard (*Xantusia vigilis*) of southwestern North America, exploits a variety of arid-land microclimates, seeking out the underside of stones and crevices and resting beneath the top surface of sand or excavating burrows. As its name suggests, the lizard emerges at night to nose about in yucca and agave plants for insects and other small creatures. As with other reptiles, the lizard is poikilothermic, enduring a state of torpor during extreme cold that causes delayed muscle contractions and even paralysis. This is when warm-blooded mammals and birds have the upper, and faster, hand.

The second-largest lizard in the world, runner-up to the well-known Komodo dragon, is the shy, retiring perentie (*Varanus giganteus*) of Australia's arid center. Also known as the Queensland goanna, it reaches about 2.5 meters (8 ft.) in length and is blotched and patched with creams, yellows, and browns, a combination that blends perfectly into the rocky, arid scrub. It saunters about by day, inspecting burrows that may contain smaller lizards, snakes, or rabbits, although it has been known to seize prey as large as juvenile kangaroos.

Another large and predatory Australian monitor lizard, up to 1.5 meters (5 ft.) long, is the sand goanna (*Varanus gouldii*). It, too, dwells in the continent's "Red Center" and can swallow a whole rabbit. When disturbed, it flees at speed on its four long legs, or even rears up on its two back legs – a tactic that gives it the alternative name of racehorse goanna. If cornered, it puffs out its body, hisses loudly, lashes its tail, and tries to bite the enemy.

Australia is home to dozens of other large lizards that are well-adapted to dry conditions. The common blue-tongued skink (*Tiliqua scincoides*) ranges across the eastern half of the continent, in arid scrub as well as woodland. It may grow to 60 centimeters (23 in.) in length and eats an adaptable diet of plant matter, including flowers and berries, and small animals such as snails and insects, as well as carrion. Rather than turn and run on its smallish, weak legs if threatened, the skink faces up to its aggressor, inflating its body, hissing loudly, and gaping its jaws to reveal its startling large blue tongue. This martial attitude is no bluff, and the skink may lunge, bite hard, and hold on while thrashing its body and tail.

The similar stumpy-tailed, or shingle-back, lizard (*Trachydosaurus rugosus*), restricted to dry regions of southern Australia, also puts on a blue-tongued defense display. It is difficult to distinguish the front and rear ends of this reptile, as both are similarly shaped, perhaps as a means of distracting or confusing predators. The stumpy tail holds a fat reserve for moisture and nourishment for use when times are harsh.

Less than half the stumpy-tailed lizard's size, at about 15 to 18 centimeters (6–7 in.) long, the Australian desert skink (*Egernia inornata*) has a copper-colored back and white bars along its flanks. It is diurnal, out by day to search for insects and other prey. At night it retreats to its burrow under a clump of spinifex or porcupine bush – one of many creatures that depend on this harsh, spiky native grass for its home.

Perhaps the most impressive display of defense among Australian desert lizards belongs to the bearded dragon (*Aphibolurus barbatus*), which reaches 60 to 70 centimeters (23–27 in.) in length. Dragons in desert areas are orange-yellow or tawny, in contrast to those in woodlands farther east, which are brown or gray. When aroused, the dragon puffs up, opens its mouth to reveal the yellow-orange interior, and erects the spiky frill of skin over its chin. It also unnervingly jumps toward the enemy.

Several lizards living in soft sandy soil show evolutionary trends to limb reduction or loss, wriggling the body from side to side to "swim" through their substrate, almost like a snake does. Only 10 centimeters (4 in.) long, the Australian two-toed desert skink (*Lerista bipes*) of the continent's dry northwest is even less lizardlike, resembling rather a scaly worm. Its front legs are absent, and its rear legs are thin and weedy, with just two toes each – one large and one small. Instead of lower eyelids, it has transparent scales that function as windows to see among soil particles as it "swims" through the sand in its search for ants and termites.

The sandfish (*Scincus philbyi*) of the Middle East matches the red-yellow color of the Arabian desert soil. It, too, wriggles like a fish but has retained large legs and toes, the latter fringed with

## DEVIL IN DISGUISE

The moloch, or thorny devil (*Moloch horridus*), is an extraordinary lizard found throughout Central Australia, including the Gibson Desert. Generally its patchy dull brown and bright orange skin matches the soils of the continent's "Red Center," although it can change its color to match the background – if not as promptly or as drastically as the champions of quick-change lizards, the chameleons. The moloch eats ants and termites, flicking out its long tongue to pick them up and then chewing the morsels in its weak jaws. Its most striking features, however, are the spines that cover its body, many of which are long and pointed like small canine teeth or rose thorns. When worried, the moloch tucks its head under its body and presents a lump of sharp, hard spikes to the enemy. Some of the spines and spiky scales also have tiny grooves that work as capillary channels, sucking up any moisture that touches the lizard's body from small puddles on the ground or leaves dampened by dew, and channeling it toward the head and mouth for drinking.

**The centralian blue-tongued skink (*Tiliqua multifasciata*) is a close relative of the common blue-tongued skink and is found in arid Central and northwest Australia.**

A northern death adder (*Acanthopus praelongus*) adopts a defensive display. Found in the deserts of northern and Central Australia, death adders are ambush predators, concealing themselves in sand or leaves and twitching the end of their tails to attract prey.

thin scales, providing extended pedal surfaces that work as sandshoes. Its long, chisel-shaped snout pushes through the soil as particles slip off its smooth, pointed tail, which is also filled with fat as a food store. This little lizard, just 20 centimeters (8 in.) long, has strong jaws for crunching hard-cased prey such as beetles and other insects. Active primarily at dawn and dusk, it sleeps by day and may hibernate in colder months. Many other desert lizards live similar burrowing lives, including the diminutive legless skinks of the Namib and Kalahari, which are less than 10 centimeters (4 in.) long and resemble not so much animals as pieces of broken-off brown shoelaces.

Only two of the world's lizards have venomous bites, and both of these are inhabitants of the arid lands of southwest and southern North America. The gila monster (*Heloderma suspectum*) grows to about 55 centimeters (21.5 in.) and warns off predators with its gaudy pink, yellow, or orange blotchy stripes. As in its 80-centimeter-long (31 in.), black-and-yellow venomous cousin, the Mexican beaded lizard (*Heloderma horridum*), the lizard's scales are domed or rounded, resembling small, shiny beads, and are adjacent rather than overlapping. The venom is made in glands in the lower jaw, and as the lizard bites and chews, it flows into the victim via the grooved front teeth. A well-fed gila monster has a torpedo-shaped tail full of fatty food stores that can last for months. Named after the Gila Basin in Arizona, this predator is mainly nocturnal and feeds on small creatures, such as insects, mice, birds, and other lizards, as well as on bird and lizard eggs. The bites of these reptiles are rarely fatal to humans but can be extremely painful.

For humans and prey alike some of the most feared desert creatures are poisonous snakes. The looping motion of the sidewinder (*Crotalus cerastes*) from North America's southwest has already been described (*see* p. 91). This agile snake grows to about 75 centimeters (29 in.) in length and hunts small mammals and lizards by night, holing up for the day in a crevice, under a bush, or in an old rodent burrow. It has a distinctive "horn" over each eye and black stripes along the top and sides of the neck that widen and break into diamond-shaped black patches along the top of the body.

The Namib Desert has its own exponent of this locomotive method, the desert sidewinding viper (*Bitis peringueyi*; *see illus.*, p. 91). Only about 25 centimeters (10 in.) in length, and with a dark brown upper body that shades to lighter brown below, it glides over the dunes with typical sidewinding skill and is just as deadly to its prey. Another sidewinding snake is the exceptionally dangerous saw-scaled viper, or saw-scaled adder (*Echis carinatus*), found in the Sahara and in deserts of the Middle and Near East. This is one of the world's most venomous creatures, and its bites kill several people each year, especially in and around the Sahara. To warn off potential predators, it coils up to rub the serrated scales on the sides of its body against each other, producing a rasping sound. In addition to the usual snake diet of small mammals, birds, and lizards, the saw-scaled viper tackles creatures that are themselves venomous, including large centipedes and scorpions.

Every desert region has a scattering of poisonous snakes, such as the snouted lancehead of the Atacama and the manushi, or Asiatic pit viper (*Agkistrodon halys*), of the Gobi. The manushi reaches 75 centimeters (29 in.) in length and ranges north from the Asian desert, across the steppes toward the great conifer forests of the northern taiga. It has rings of dark gray, brown, or black alternating with light yellow, cream, or fawn, and eats mainly small mammals. It hibernates from about October to March, due to the cold in the north of its range and the aridity in the south.

Some snake surveys point to Australia as home to eight of the world's ten most poisonous land snakes. One of these, the fierce snake (*Parademansia microlepidota*), is very large – at 2.5 meters (72 in.) in length – very fast, and very venomous, and inhabits the arid zones in the center–east of the continent. The eastern brown snake (*Pseudonaja textilis*) is another sizeable species, at 1.5 meters long (60 in.), and ranges into the continent's eastern deserts. It is an efficient predator of small animals. When it strikes, it does not jab in its poison and then retreat, but hangs on with its fangs so as to inject additional venom. Less dangerous to humans is the bandy-bandy (*Vermicella annulata*), a snake of about 80 centimeters (30 in.) long with alternating black-and-white hoops along its body. It hunts smaller snakes and lizards and, for a top carnivore, is relatively common throughout the Central Australian desert regions.

All of these snakes benefit from their carnivorous diet, which usually contains plenty of moisture in the form of blood and other body fluids. Like other reptiles, they use the heat retained by the desert soil to incubate their eggs, which in nearly all cases are buried by the mother and then abandoned.

### Amphibians of the desert

Frogs, toads, and other amphibians usually need water in which to spawn – for the first stage of their lives tadpoles are aquatic gill-breathers. Nevertheless, surprisingly, a few amphibians have managed to survive in the desert environment.

In North America various species of spadefoot toads (genus *Scaphiopus*), such as the western spadefoot (*S. hammondi*), are well adapted to desert life. They are mostly mottled pale green and brown in color, dig burrows with their club-shaped back feet for shelter by day, and emerge only at dusk to hunt for insects and spiders. As the dry season takes grip, they extend the burrows to 1 meter (39 in.) or deeper and stay there all the time – in rare cases for as long as six months. During this time they live off nutrient reserves in the body and may lose up to half their body weight. They also store virtually pure water rather than urine in the bladder, which may swell like a balloon to take up half of the toad's volume, while the waste substance urea generally found in urine is instead retained in concentrated form in the blood and other tissue fluids.

Surrounded by damp soil in its deep burrow, the toad acts like a sponge and draws water into itself from the soil by a natural process of osmosis (whereby dissolved substances exchange solutes or solvents to even out their concentrations). When the brief rains arrive, the toads emerge, mate, and lay eggs quickly in the puddles. The whole life cycle from fresh spawn to dry-land toadlet lasts just six weeks.

## THE "RED RACER"

Perhaps the most commonly sighted snake of arid regions of North America is the coachwhip (*Masticophis*, sometimes *Coluber, flagellum*), which is often seen sunning itself on desert roads and tracks in the early-to-late morning hours. On average this long, slender whiplike snake is some 1.2 meters (4 ft.) long – though it can be as much as twice that length – and has a tail marked like a plaited whip.

The coachwhip's dorsal coloration is predominantly tan or gray to reddish, with a pinkish underbody, a light-brown head, and a black neck with thin white crossbars. The eastern subspecies tends to be brownish, while western subspecies tend to be reddish. As the western subspecies gets older, it begins to take on a more distinct reddish appearance.

The coachwhip is a swift and aggressive hunter, and is perhaps the fastest snake in the Americas – perhaps occasionally moving at 11 kilometers per hour (7 mph) – giving it its popular name of "red" or "black racer." It preys on lizards, small mammals, large insects, and occasionally even young rattlesnakes, killing by biting while the prey is pinned under its coils. In defense it bites with a jerk of the head, causing a torn wound. It is nonvenomous and, despite its aggressive behavior, is sometimes successfully kept as a pet.

A male ostrich (*Struthio camelus*) guards its young and unhatched eggs in the Sahara. Sitting on the nest keeps the eggs cool during the daytime and warm at night.

The tough, hardy green toad (*Bufo viridis*) is often found in the dry, sandy soils of North Africa and Central Asia, as well as elsewhere. Like most toads its warty skin produces both a layer of slimy mucus to cut down water loss and also distasteful or even poisonous secretions to deter enemies. The female green toad is a champion egg-layer among amphibians, producing up to 20,000 eggs. This excess helps counteract the precarious conditions in which such toads live, as seasonal puddles and pools soon dry out.

The water-holding frog (*Cyclorana platycephalus*) of Australia spends a large percentage of its life buried about 50 to 100 centimeters (20–39 in.) below the desert's surface, in a self-generated water chamber. Having burrowed beneath the desert surface, the frog loosens its thin outer layer of skin and mucus to form a cocoon that retains water between it and the frog, almost like a water-filled sleeping bag. The body also stores copious water in its tissues and bladder. Large and stout with green-gray blotches, the water-holding frog maintains its underground state until it detects a rise in the moisture content of the surrounding soil, a sign of recent rainfall, at which point it hurriedly digs up to the rain-dampened surface with its shovel-like feet to mate and lay eggs in temporary pools. The tadpoles grow up to five times faster than ordinary frog or toad larvae and are ready to leave the water as miniature adults after just two weeks, rather than the more usual six to ten. Meanwhile the mature adults feed hungrily on insects and other small creatures that have been encouraged by the rainy spell. As drought grips the land again, the frogs, their bodies loaded with water, disappear below the surface once more.

## Birds of the desert

With generally higher body temperatures than mammals, birds are better equipped for survival in the searing desert heat. Most birds, too, are able to fly long distances in order to find water, or to move on if a dry spell becomes too acute. Consequently many bird species are found on the fringes of, though rarely deep inside, true deserts; they include the world's two largest birds, as well as some of the smallest.

The ostrich (*Struthio camelus*) stands over 2.5 meters (8 ft.) tall and is not only the biggest living bird but also the fastest creature on two legs, striding with enormous paces, rather than running, at more than 70 kilometers per hour (43 mph). It is ideally suited to grasslands and arid patches across Africa, including the fringes of the Sahara. Its wings are useless for flight but their soft plumes provide shade for the lower body, especially the almost naked patches of skin just beneath the wings where excess body heat radiates into the air. Ostriches are desert nomads, wandering wherever the rains encourage plant growth so that they can feed on shoots, buds, leaves, seeds, and other vegetation. However, in the tradition of true desert survivors, they are not fussy and also peck up insects, little lizards, mice, and other small animals.

When conditions are suitable, the male ostrich rounds up a harem of several females. The chief

**The tiny, insectivorous elf owl (*Micrathene whitneyi*) often roosts and nests in giant cacti, such as the saguaro, of Mexico and the American Southwest.**

female scrapes a shallow nest and lays her eggs, and the others may add to their number. The eggs are gigantic, each one up to 15 centimeters (6 in.) long and the equivalent of 30 to 40 hen's eggs in volume. The chief female sits on them by day, more often to shade them from excessive heat rather than to keep them warm, while the male takes over incubation at night. The eggs hatch after 40 or so days, and the young birds can run and adopt a nomadic lifestyle almost immediately.

Second in size only to the ostrich, the emu (*Dromaius novaehollandiae*) of Australia also speeds across the desert with huge 3-meter (1-ft.) strides, and is able to sprint at 60 kilometers per hour (37 mph) for short bursts. Its diet is even wider than the ostrich's, comprising seeds, fruits, leaves, insects, worms, farm crops, and carrion. The female lays eight to ten greenish-brown eggs in a nest 1 meter (40 in.) across, under a tree or bush, while the male is both the nest's constructor and the major incubator, sitting on the eggs for two months. When extreme drought forces emus from the outback, they can invade farmland and become serious pests. The third-largest birds are the rheas (*Rhea americana* and others) of South America. They, too, are wanderers of grassland, arid scrub, and open country and have a similar diet to the ostrich.

At the other end of the size spectrum from the emu, arid Australia is also home to the budgerigar (*Melopsittacus undulatus*), one of the smallest members of the parrot family. "Budgies" of all colors are familiar as cage birds around the world, but those of the outback are predominantly green. Noisy and gregarious, they drink early in the morning at a local pool or wait until a kangaroo or similar large animal has scraped a hole in the soil to get at water. Having quenched their thirst, they take off as a whirling, wheeling, chattering flock for a morning feed of grass seeds. After roosting in the shade during the heat of midday, they feed again toward evening and then settle for their night's rest.

As with emus and ostriches, budgerigars are nomadic. However, during their rainfall-triggered breeding season, they stay put as a colony for a few weeks, nesting in holes in tree trunks, old logs, and rock crevices. The female incubates her four to seven eggs for about 18 days, and the offspring are ready to fly after four weeks.

Another desert mini-resident is the elf owl (*Micrathene whitneyi*) of southwest North America, which measures only about 13 centimeters (5 in.) long from its beak to the tip of its very short tail – about the size of a human hand. It roosts by day in a tree, bush, or tall cactus, such as the saguaro. At night it is a predator of flying insects such as moths, beetles, grasshoppers, and crickets. Unusually for an owl, it can hover to snatch victims from foliage, and it also swoops to the ground for small lizards, snakes, and scorpions – hastily nipping off the poisonous sting. The elf owl's usual nesting site is a hole in a tree or large cactus, perhaps excavated by a gila woodpecker. The female incubates her two to five eggs for about 25 days and the male feeds both her and the chicks when they hatch.

Since lizards and snakes are relatively common in many desert regions, several birds habitually feed on them. In the Sonoran, Mojave, and nearby areas of North America the greater roadrunner (*Geococcyx californiana*), a member of the cuckoo family, resembles a rooster with a dark head crest and long, straight tail. It can run at speed, up to 25 kilometers per hour (16 mph), preferring to use its legs rather than its wings. The roadrunner feeds on a range of small creatures, including grasshoppers, crickets, and other insects, but also birds' eggs, lizards, and snakes – even poisonous varieties such as the rattlesnake. Roadrunners pair for life and raise their three to six chicks in a bush or clump of cacti.

In African deserts south of the Sahara the secretary bird (*Sagittarius serpentarius*) stalks the dry bush, on the lookout for snakes and lizards as well as small mammals, birds, and insects. This tall, stately bird has very long legs and can stride more than 30 kilometers (19 mi.) on a day's foraging. Like the roadrunner, it kills prey with a very swift, hard peck, usually to the neck region, or it may pin down its victim with its large, strong, heavily scaled feet or even stamp it to death. In some regions the secretary bird is protected by laws or local traditions because of the toll it takes of poisonous snakes. The species is named from the head crest of several long feathers that protrude upward and back, like the quill pens that a human scribe of previous centuries might have put behind his or her ear.

Several larger birds of prey range huge distances over arid grassland, scrub, and desert in search of victims. The lanner falcon (*Falco biarmicus*) soars across dry areas of Africa south of the Sahara, including the Kalahari Desert, and over the Middle East, and is rarely found in areas with annual rainfall of more than 60 centimeters (23 in.). This falcon is a majestic hunter, with a wingspan of almost 1 meter (3 ft.), its black forehead and "moustache" contrasting with the white cheeks and throat and blue-brown back with creamy-buff underparts. Unusually for falcons, it often attacks prey on the ground or just above, taking small and medium-sized birds, rodents, bats, and lizards.

In southern North America and South America the smaller, colorful, and adaptable aplomado falcon (*Falco femoralis*) is a species found in dry, open lands characterized by grasses, cacti, and mesquites. Like many arid-land birds, it sweeps to and fro near bushfires for insects and small birds that have been disturbed by the flames.

In Australia one of the world's largest birds of prey, the wedge-tailed eagle (*Aquila audax*), is a familiar sight over the dry outback. It is glossy brown-black and measures 1 meter (3 ft.) from beak to wedge-shaped tail – its distinguishing feature as it soars in search of rabbits, hares, small wallabies, and young kangaroos. Like many other birds of prey,

## AERIAL WATER

Sandgrouse, which are not close relatives of true grouse, live in dry lands and deserts from southern Europe and North Africa across to Asia. They resemble smallish pigeons and are sturdy, mainly sandy-colored, fast-flying birds with long wings. Several species are famed for their habit of flying dozens of kilometers to a water source in the breeding season, soaking up the precious moisture in their specially adapted breast feathers, and returning with this aerial cargo to the chicks in the nest, who lap and suck up the water. Here a flock of Burchell's, or spotted, sandgrouse (*Pterocles Burchelli*) land at a waterhole near the Nossob River in the Kalahari-Gemsbok National Park in South Africa.

**One mammalian survivor of arid regions is the desert hedgehog (*Paraechinus aethropicus*), at just 14 to 28 centimeters (5.5–11 in.) long the smallest of all the hedgehog species. Found in the deserts of northern Africa and the Arabian Peninsula, desert hedgehogs live in underground burrows or in rocks and cliffs.**

it also takes carrion and is sometimes persecuted by humans for killing lambs when it is in reality only scavenging on a dying or already dead carcass. Another well-known avian scavenger throughout the Americas is the turkey vulture (*Cathartes aura*), with its mainly black plumage, bare head and neck, purplish face, and white beak. It ranges from deserts to forests and gathers in considerable numbers at a large carcass, or feeds on birds' eggs, chicks, and even old fruit.

## Smaller mammalian herbivores

The mammalian fauna of most deserts is dominated by small rodents, including hundreds of species of ground squirrels, mice, rats, kangaroo mice, kangaroo rats, gerbils, jerboas, and jirds. Their burrowing, nocturnal habits are described earlier. One of the best-known, from its success as a small, clean, fairly odor-free pet, is the Mongolian gerbil – technically called the Mongolian jird (*Meriones unguiculatus*) – of the Gobi Desert and surrounding regions. This gerbil exhibits a wide array of adaptations to desert life. It emerges at night to seek dew-dampened seeds and other plant foods and so never needs to drink. It may take food back to its burrow to eat it there, out of danger, or to store it for later. One gerbil is recorded to have hoarded 20 kilograms (44 lb.) of assorted seeds in its underground home. Stored seeds in the burrow also act as sponges, soaking up water from the humid air and moist soil to provide the gerbil with extra moisture. The bulk of each seed is carbohydrate (plant starch), which yields metabolic water when digested and broken down inside the body.

The gerbil's pale brownish upper fur gives good camouflage on the desert soil, and the large eyes, big ears, and long whiskers are adapted for night activity. In common with many other small desert rodents, the large back feet can hop at speed across soft, open ground. The feet are furry underneath, both to aid grip and as insulation from the hot sand. The long tail gives balance and works as a rudder when bounding at speed, and even the gerbil's underside is white to reflect heat from the ground below.

In the Atacama and Patagonian deserts of South America the peludos, or hairy armadillos (genus *Chaetophractus*), make burrows over 1 meter deep (3 ft.) and several meters long on the slopes of sand dunes. They are diurnal in winter, when more than half of their diet consists of plant food. However, they become more nocturnal in summer to avoid the higher desert temperatures and eat more small animals such as insects, lizards, snakes, and mice. Like other armadillos, they are well protected by horny crosswise plates and, when threatened, roll into a ball. The smallest armadillo species, able to fit on a human palm, is the fairy armadillo (*Clamyphorus truncatus*) of Patagonia. Like other armadillos, it is a powerful digger, with large claws on its strong forelegs. It eats plants, ants, worms, and snails. However, plowing up the scrubland for farming and attacks by domestic dogs have gradually reduced its range to the desert regions of west and central Argentina.

### The larger herbivores

Many arid regions are now home to livestock such as cattle, sheep, and goats, as well as those most famed of large desert-dwellers, the camels. Various antelopes, gazelles, and ibex thrive on sparse grazing in semidesert regions of Africa and Asia. Probably the best adapted for true desert life are the dama and dorcas gazelles (*Gazella dama* and *G. dorcas*) of the Sahara. They are slender and long-legged, pale fawn on the upper parts and white underneath, with S-shaped, annular horns. They move with the rains to find the greenest vegetation, eating grasses, herbs, and woody plants. In the breeding season each male defends a patch of territory from other males and gathers a harem of females for mating.

The pronghorn (*Antilocarpa americana*) of North American deserts is sometimes called an "antelope," but it is in fact more closely related to cattle and deer. It has a small, forward-pointing branch, or "prong," on each main horn and horizontal dark and light stripes down its neck. It is one of the fastest of all land animals, running at 65 kilometers per hour (40 mph), and is able to maintain a speed of 40 to 50 kilometers per hour (25–30 mph) for many minutes, enabling it to outpace most predators. It ranges across prairies and semidesert, existing on tough vegetation and obtaining most of its water from its food. Like most desert-dwellers, it produces concentrated urine, thereby losing little body water.

In South America the main large herbivores are members of the camel group. They include the guanaco (*Lama guanicoe*), which inhabits high regions of the Patagonian Desert, and the vicuna (*Vicugna vicugna*) on the Andean mountain edges of the Sechura and Atacama deserts. They resemble smaller, slimmer, more dainty versions of their close and more familiar cousin, the llama. Indeed, the guanaco is probably the wild ancestor of domesticated llamas and alpacas. Both species feed on various forms of vegetation, especially grasses, and have very long, woolly fur to keep out the mountain's nighttime cold and daytime heat.

The wild asses of Africa and Asia also wander in semidesert and rugged country. They can withstand great heat and move nimbly among the rocks to find food and shade. The African ass (*Equus africanus*), wild ancestor of the domesticated ass, has suffered from military conflict through much of its range.

### The desert's top predators

In any habitat the large carnivores or top predators are relatively few in number. This is a consequence of the natural ecological balance between species. In order to allow prey populations to maintain themselves, and to avoid hunting them to extermination, the top carnivores need to range over a wide area and take one meal here, another there some days later, and so on. In a rain forest, where there is a plethora of densely packed prey, a big cat, such as a jaguar, may occupy a hunting territory of less than 5 square kilometers (2 sq. mi.). In desert regions, by contrast, large meals are so sparse that the most dry-adapted of big cats, the cheetah of Africa and – possibly still – the Middle East, may range over an area greater than 500 square kilometers (195 sq. mi.). This compares to as little as 20 square kilometers (8 sq. mi.) for a cheetah living in productive savanna.

The food of a carnivore usually contains plenty of fluid – as blood and other body liquids – compared to the dry seeds of herbivores, so obtaining sufficient water tends to be less of a problem. Many mammal predators, such as cheetahs and other cats as well as members of the dog group such as hyenas and foxes, mark their territories with scent and urine and usually chase away any intruders they

## ENDANGERED ORYX

Oryx were once among the most successful larger mammals to colonize desert regions of Africa and the Middle East. During the 20th century, however, increased hunting of the Saharan scimitar oryx (*Oryx dammah*) and the Arabian oryx (*Oryx leucoryx*) of the Middle East left both species critically threatened.

Oryx scrape holes in sand dunes with their hooves and horns to expose cooler ground and make a depression shaded from the sun. They graze on grasses and shrubs in the cooler periods after dawn and before dusk. In the absence of water they dig up bulbs and other succulent plant parts and can survive on the moisture in these for weeks without taking a drink. They can also detect rain from many kilometers away and trek to the region in order to find fresh plant growth.

The last Arabian oryx in the wild were sighted in the 1970s. However, from the 1950s onward small herds of oryx were taken into captivity as part of a conscious effort to save the species. Reintroductions to the wild were made in 1982 in Oman, 1983 in Jordan, and 1990 in Saudi Arabia. These appear to have been partially successful and some 500 Arabian oryx now roam their original desert lands, with some 300 still held captive for breeding in the Middle East and 2,000 elsewhere, mainly in the United States.

encounter. In the grip of drought, however, the energy required to defend a patch often exceeds the nourishment and moisture available and territorial boundaries break down. Some normally lone-hunting species, such as brown hyenas (*Hyaena brunnea*), may even band together on communal forays during this temporary truce.

The survival of big cats in arid, treeless places is somewhat surprising because the cat group originated primarily as tree-climbing forest hunters. However, the cheetah (*Acinonyx jubatus*) hardly ever climbs, unless it is on to a termite mound to scan the surroundings. It is the only cat with non-retractable claws, and these are larger and blunter for good grip on the earth. The claws are just one of the cheetah's adaptations for its bursts of speed over open ground. Others include its lean, rangy build, with long, slim legs and a flexible backbone that flips up and down to increase stride length.

Cheetahs drink on average once every three to five days and can survive for more than 10 days without water. They are supreme sprinters, being the fastest creatures on legs with a top speed of perhaps as much as 100 kilometers per hour (62 mph). This lightning dash is brought into play after the cheetah has stalked and crept as close as possible to its prey, usually a small antelope or gazelle, such as a wildebeest calf or Thomson's gazelle, or perhaps an African hare. When within 50 meters (55 yd.) of its prey, the cat accelerates to top speed in less than two seconds. However, it can maintain maximum velocity only for about 500

A pair of cheetahs (*Acinonyx jubatu*) drink from a pool in the Kalahari-Gemsbok National Park, South Africa. The cheetah is virtually extinct outside Africa, with only a few scattered populations in Iran and Russia. Estimates are that there are fewer than 9,000 cheetahs left on earth.

meters (550 yd.) or 20 seconds, after which it tires rapidly. The average chase occurs across 160 to 180 meters (175–200 yd.), and the cheetah has a 40- to 55-percent chance of a successful catch.

Cheetahs were once widespread across Africa and the Middle East to India. Now they are almost extinct in Asia, apart from possible hangers-on in remote desert areas such as the Kara-Kum and Iranian Desert, where they are largely free from human persecution. Even in their African stronghold cheetahs are declining in the north and west, driven to desert edges where they sometimes run into trouble when hunting livestock.

Several other cats have become adapted to life in dry lands. The sand cat (*Felis margarita*; *see illus.*, p. 92) of the Sahara and Middle East is hardly larger than its domestic relative. Its yellow-brown or gray-brown coat blends perfectly with the desert soil as it preys on rodents, lizards, and large insects. The caracal (*Felis caracal*) has a similar distribution but is much larger, about 1 meter (3 ft.) long from nose to tail. Its tufted ears have given it the alternative names of African or desert lynx. With its slim, long-legged build, the caracal runs down hares, rabbits, large rodents, and small-hoofed mammals such as baby deer, antelopes, and gazelles. It also leaps vertically, springing straight up from a standing position to "bat" down low-flying birds with its paws.

In North America the bobcat (*Felis rufus*) and puma, or cougar or mountain lion (*Felis concolor*), are adaptable cats that are sometimes driven into semidesert regions by human persecution. Both hunt smaller prey such as jackrabbits and young mule deer; the puma also takes adult deer and, occasionally, farm livestock. In South America the margay, or tigrillo ("little tiger"; *Felis wiedi*) also frequents dry scrub around the fringes of true desert in its search for rats, squirrels, and similar rodents as well as various kinds of birds.

## Desert dogs and foxes

Two members of the dog group that survive in varied habitats, including dry grassland and scrubby semidesert, are the coyote (*Canis latrans*) and the dingo (*Canis dingo*) of Southeast Asia and Australasia. The latter is tawny yellow in color and can hardly bark, only howl. Sometimes regarded as a subspecies or breed of domestic dog (*Canis*

A family of dingoes sleeps in the sun in the Australian outback. The family group is the basic unit of the dingo's social organization.

*familiaris*), it may have accompanied the first human settlers to reach Australia, some 40,000 years ago, perhaps as a part-domesticated companion for hunting and protection. Since then dingoes have become semiwild or wild and live in bush and semidesert regions, in family groups. They prey mainly on rabbits, wallabies, small birds, lizards, and snakes, but have sometimes come into conflict with ranchers due to their occasional sheep-stealing habits.

The coyote's mournful howl echoes through the night in the arid, scrubby southwest of North America and across most of the rest of the continent as well. In dry regions, where prey is small and sparse – such as ground squirrels, rabbits, and mice – these dogs tend to live alone or in pairs. In woodlands, where meat is more plentiful, they may form larger packs. Like the dingo and many other predators, the pressure of persecution has driven coyotes farther into drier areas, where they receive less human attention.

Two species of foxes that are both suited to desert life are the kit fox (*Vulpes macrotis*) of North America and the fennec fox (*Vulpes zerda* or *Fennecus zerda*) of North Africa, including the deep Sahara. Both species have huge ears, both to hear their prey in the desert night and to work like radiators and rid the body of excess internal heat. The fennec is the smallest of all foxes, only 60 centimeters (23 in.) from nose to tail, but its ears may be 15 centimeters (6 in.) tall. It hides in a burrow by day and emerges at dusk to hunt small prey such as mice and beetles. The kit fox is slightly larger and preys on rabbits, hares, and kangaroo rats.

**The fennec fox (*Vulpes* or *Fennecus zerda*) has an unusual feature for desert creatures – its thick, fine fur, which insulates the fennec from the cold nights as well as the daytime heat. The feet are also thickly furred to insulate the pads from the hot desert sand and to cushion the fennec's step, which makes its movement quiet and difficult to detect by prey.**

**A Merriam kangaroo rat (*Dipodomys merriami*), photographed in the Kofa National Wildlife Refuge, Arizona. The rats have large hind feet that make it easier to hop across hot sand.**

A much smaller but equally fierce hunter of the American Southwest is the ringtail (*Bassariscus astutus*), a weasel-like member of the racoon family with a bushy, ringed tail. It climbs well, frequents rocky areas, and hunts alone at night for insects, mice, small birds, and similar prey. However, it is an opportunist and also feasts on fruits, buds, and other plant matter.

### Hyenas and jackals

Large animals often die in the desert – of thirst, hunger, heat stress, or all three. There are always scavengers, however, to feed off the flesh and crunch the bones. Of the three main species of hyena, best adapted to very dry habitats is the brown hyena (*Hyaena brunnea*). Hyenas are much more actively predatory than their popular image would suggest and are ready to snap up small living prey such as insects and mice, raid birds' nests, and chew fruits and berries. One night, one brown hyena in the Kalahari was observed to remove and cache all 26 eggs from an ostrich nest, which it recovered and ate over the next two days. Along the desolate shores of the Namib Desert these omnivores scavenge on dead seabirds, washed-up fish, stranded whales, sick seals, and almost anything else the Atlantic Ocean can throw up – hence its alternative names of beach wolf and strand wolf.

The golden and silverbacked jackals (*Canis aureus* and *C. mesomelas*) are doglike predators of dry grasslands and arid scrub. The former ranges around eastern Europe, North and East Africa, and across to India; the latter is chiefly found in East and South Africa. Like the hyenas, and contrary to popular opinion, they depend less on scavenging and more on opportunistic hunting for small animals such as rodents and insects, plus foraging for fruits and other vegetable foods. Jackal family life is remarkably stable. The male and female pair-bond for many years, and the young stay with them for the first few years. Juveniles from elsewhere may also work their way into the clan so that the pack can hunt cooperatively to bring down larger-hoofed mammals such as gazelles.

### Australia's desert marsupials

The earth's driest continent, Australia, is also home to dozens of marsupials – mammals whose young are born at a very early stage of development, and which continue their growth outside the womb in a body pouch or pocket (marsupium), usually found on the belly of the female. Several marsupial species are highly adapted to the arid interior of the country and follow lifestyles equivalent to the nonmarsupial, or placental, mammals, in other desert regions.

The marsupial mole (*Notoryctes typhlops*), which is found mainly in southwest Australia, spends its life tunneling through soft sand in search of grubs, worms, and other soil creatures. The mole tunnels about 8 to 10 centimeters (3–4 in.) below the surface, using its hard nose pad and the massive spadelike claws on its front feet to push through the soil. Since the usual marsupial front-opening pouch for the young would soon fill with sand as the mole digs underground, this species has evolved a rearward-opening version.

The rabbit-eared bandicoot (*Macrotis lagotis*) is an inhabitant of scattered wood and arid scrub in Central and northwest Australia. It resembles a long-nosed rat but possesses large kangaroolike rear feet for hopping and huge rabbit-type ears. Such outsized ears are a common feature of many mammalian desert-dwellers – from jackrabbits and fennec foxes to wild asses. They catch the faintest sounds of prey or predators and also help to radiate excess body heat into the air.

Large ears are one of many desert-adapted features of the biggest marsupial, the red kangaroo (*Macropus rufus*). The reddish-colored male may grow to 1.8 meters (6 ft.) tall and is known locally as a "boomer"; the female, by contrast, is pale blue-gray and called a "blue flier." Red kangaroos live in small family parties of four to eight members and are the most desert-dwelling of the kangaroos and wallabies, lazing in the shade of trees and termite mounds. They smear saliva on their bellies, feet, and tail to cool by evaporation and become active in the evening as they consume grass, leaves, roots, and soft bark. Their bouncy, hopping gait is ideally suited to open country with soft soil – a big individual can bound along at about 50 kilometers per hour (30 mph), with massive single hops some 10 meters (33 ft.) long and 3 meters (2.5 ft.) high.

GREAT BASIN
MOJAVE/
SONORAN
NORTH ATLANTIC
OCEAN
SECHURA
ATACAMA
PATAGONIA
SOUTH ATLANTIC
OCEAN
PACIFIC OCEAN

# AMERICAN DESERTS

*The arid regions of the Americas encompass an extraordinary range of landscapes, from the gaping canyons and blistering salt flats of the Great Basin to the cacti forests of the Sonoran and the windswept plains of Patagonia. Many of the deserts lie beyond the world's subtropical arid zones, maintained instead by the rain-shadow cast by the Americas' mountainous "backbone" or by cold offshore currents.*

A cactus forest sprawls across the Anzo Borrego Desert – part of the Sonoran Desert, east of San Diego.

# Great Basin

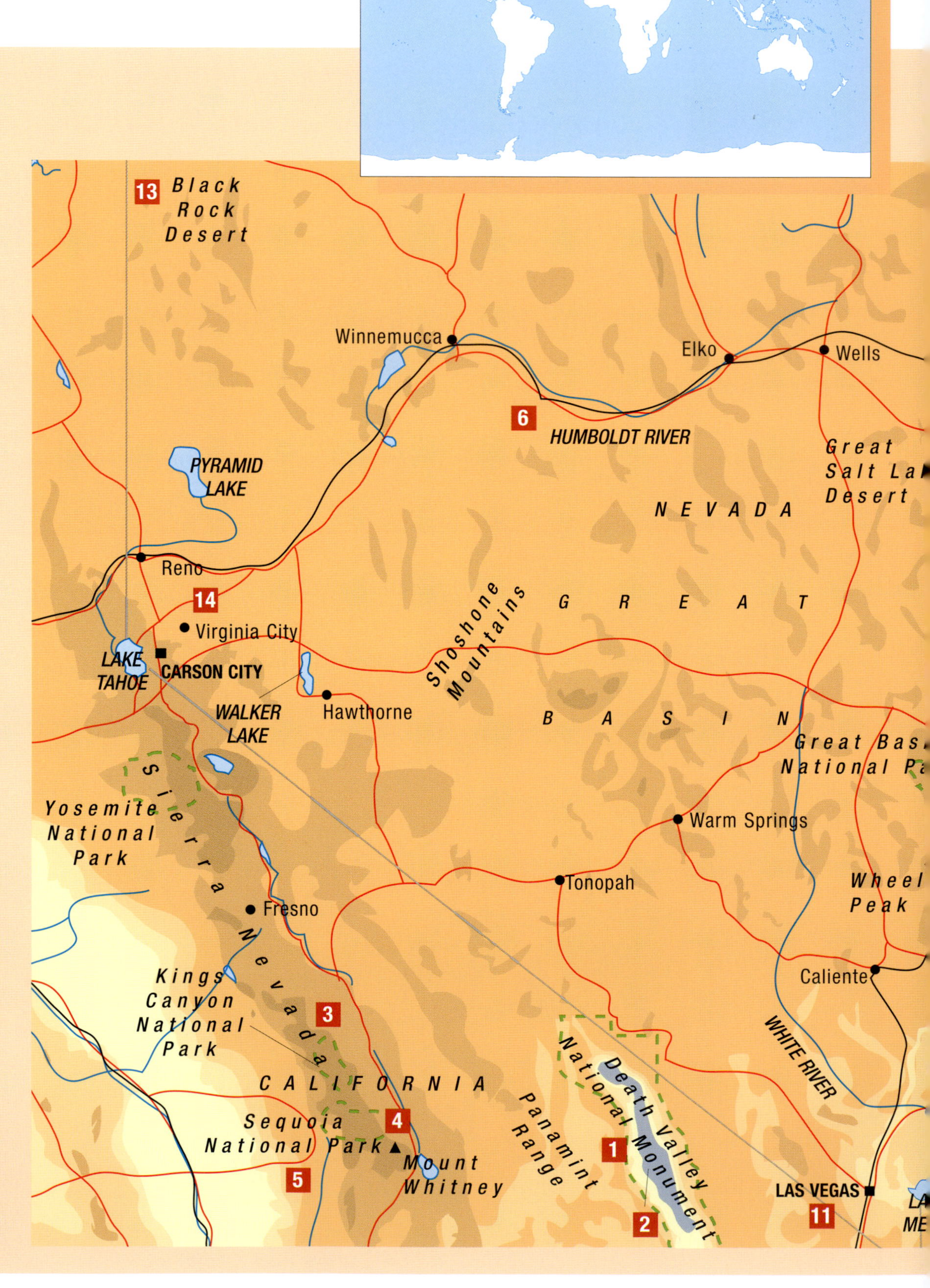


**1. Death Valley**
The hottest, driest, and lowest place in North America is Death Valley, eastern California, where temperatures can soar to more than 50ºC (122ºF) in summer (*see* p. 20 *and illus.*). The site was established as the Death Valley National Monument in 1933.

**2. Badwater**
This small salt pool is the only home of the Death Valley snail. About 8 kilometers (5 mi.) west of Badwater is the lowest point on land in North America – 86 meters (282 ft.) below sea level.

**3. Sierra Nevada**
The western boundary of the Great Basin is the spectacular Sierra Nevada mountain range. The mountains block rain-bearing winds from the Pacific, casting a rain-shadow over the entire region.

**4. Mount Whitney**
At 4,418 meters (14,494 ft.), Mount Whitney in the Sierra Nevada is the United States' tallest mountain outside Alaska.

**5. Sequoia National Park**
The magnificent giant sequoias of Sequoia National Park in California are among the world's oldest and tallest trees. One specimen, "General Sherman," is 83 meters (272 ft.) high and is thought to be 3,000–4,000 years old.

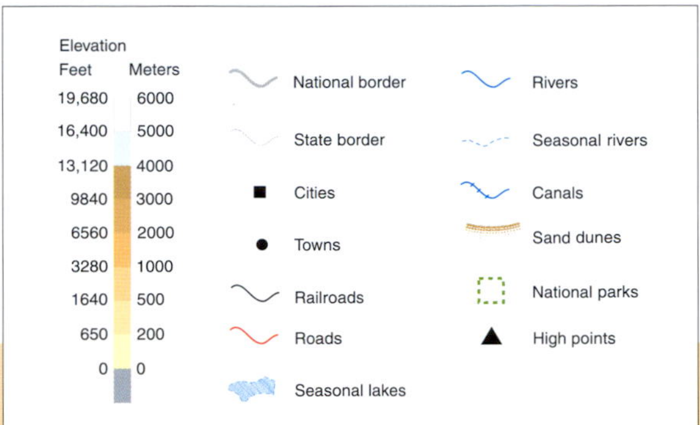


**6. Humboldt River**
At 467 kilometers (290 mi.) long, this is the Great Basin's largest perennial watercourse. Rising northeast of Elko, it flows into the Humboldt Sink without reaching the sea.

**7. Great Salt Lake**
One of the natural wonders of the world, the Great Salt Lake in northern Utah is eight times saltier than the sea. It is fed by freshwater streams but has no outlet, so its waters continually evaporate in the heat, leaving salt behind. The first nonnative to visit the lake was the fur trader James Bridger in 1832. The arid land to the southwest of the lake is the Great Salt Lake Desert.

**8. Great Salt Lake Desert**
Much of this flat, low barren area, some 175 kilometers (110 mi.) long, is used by the military and is off-limits.

**9. Great Basin National Park**
This park incorporates the scenic Snake Mountains. The tall peaks rising from the desert floor catch enough snow and rain to support forests. Among the trees are bristlecone pines – the oldest trees in the world (*see* p. 82 *and illus.*).

**10. Lehman Caves**
A honeycomb of limestone caverns full of stalagmites and stalactites lies under the eastern slope of Wheeler Peak in Great Basin National Park. The caves were used as burial chambers by prehistoric people.

**11. Las Vegas**
This city of 158,295 people in the southeastern corner of Nevada is an oasis of commerce in the desolate Great Basin desert. Relaxed gambling laws, extravagant casinos, and luxurious theme hotels attract an estimated 30 million visitors every year.

**12. Grand Canyon**
This gorge on the Colorado River where it flows across the northwest corner of Colorado is some 451 kilometers (280 mi.) long, 6 to 29 kilometers (4–18 mi.) wide, and in places 1.6 kilometers (1 mi.) deep. It was established as a national park in 1919.

**13. Black Rock Desert**
One of the largest sinks in the Great Basin is Black Rock Desert, a salt flat some 110 kilometers (70 mi.) long and 32 kilometers (20 mi.) wide. It is a famous site for land-speed racing.

**14. Comstock Lode**
In 1859 a rich vein of gold and silver called the Comstock Lode was discovered near Virginia City, causing a boom in the local economy and population. Many of the gold-rush settlements in the area are now ghost towns.

# Fact File

- The Great Basin covers about 540,000 square kilometers (208,000 sq. mi.) and includes most of Nevada, as well as parts of Utah, Oregon, Idaho, Wyoming, and California. It is bound by the Sierra Nevada mountains in the west and the Wasatch Range (part of the Rockies) in the east. The southern part of the basin includes the Mojave Desert (*see* pp. 118–121).

- Despite its name, the Great Basin is not a single basin but a vast area of mountains and valleys without drainage to the sea. Watercourses in the basin drain into low-lying desert flats, called sinks, in the valleys. Some of these contain salty water, but others are seasonally or nearly permanently dry.

- The Sierra Nevada mountains are responsible for the region's arid climate. The mountains cause humid air from the Pacific to rise and cool, which causes moisture to precipitate as snow and rain over the western flanks of the mountains. As a result, air flowing eastward beyond the mountains is dry.

- The dominant plant in the Great Basin's arid valleys is a silvery gray shrub called sagebrush. Mid elevations support low, shrublike trees such as piñon and juniper, while the mountains are covered by forests of ponderosa pine, Douglas fir, and other conifers. Sagebrush is the state flower of Nevada.

- Mountain animals of the Great Basin include mule deer, beavers, muskrats, and porcupines, as well as predators such as bobcats, lynx, foxes, and cougars. Small nocturnal animals, such as kangaroo rats and pack rats, are more common in the desert areas, as well as ground squirrels, coyotes, skunks, and numerous reptiles, including desert tortoises, rattlesnakes, and gila monsters.

- Around 15,000 years ago the climate in the Great Basin was much cooler and wetter. Forests of willow, oak, and Monterey pine flourished in the valleys, and the wildlife included mammoths, sabertooth cats, and giant ground sloths.

- Two vast lakes – Lake Bonneville and Lake Lahontan – once filled much of the northern part of the Great Basin. When the Ice Age ended and the climate got hotter, the lakes dried up. By about 10,000 years ago all that remained was Great Salt Lake (a remnant of Lake Bonneville) and a few pools of brackish water.

- Black Rock Desert is a famous site for land-speed racing. The salty ground of this prehistoric lake bed is almost perfectly flat and becomes as hard as concrete by the end of summer. The current land speed record of 1,233.7 kilometers per hour (766.6 mph) was reached at Black Rock Desert in 1997.

## Valleys and Mountains

Few parts of the world are as rich in dramatic scenery and climatic extremes as the Great Basin of North America. From the snowy peaks of the Sierra Nevada at the basin's western edge, the land plunges down through picturesque valleys such as Yosemite and Lake Tahoe into an arid wilderness of scrubland, punctuated by rugged mountain ranges. In the south lies scorching Death Valley, an alien world of sand dunes and moonlike rock formations, while in the southeast is the Grand Canyon, where the world splits open to reveal a chasm of breathtaking proportions. To the north are the great salt flats of Black Rock Desert and Great Salt Lake Desert – the salt-encrusted beds of prehistoric lakes that dried up after the Ice Age.

The Great Basin is North America's largest arid and semiarid region, covering around 540,000 square kilometers (208,000 sq. mi.). The Sierra Nevada mountains and the Rocky Mountains (the Wasatch Range) define its western and eastern boundaries; the north is bordered by the Columbia Plateau; and the southern part of the Great Basin merges into the Mojave and Sonoran deserts.

The Great Basin is neither a single basin nor, strictly speaking, a true desert. Only the valleys have a desert climate, but even here there is enough water from the occasional storm or from meltwater streams to support vegetation. Sagebrush is the most common plant, often dominating the landscape for miles on end; there are very few cacti and only occasional yuccas. The mountains rising from the desert floor are tall enough to catch what little moisture blows over the Sierra Nevada, and many are covered with pine forests and capped with snow.

## PEOPLE OF THE DESERT

Before the arrival of European colonizers, the Great Basin was home to a small population of nomadic native Americans, who were forced by the harsh climate to live as hunter-gatherers. After Europeans colonized the Americas, the basin was a barrier to travel, deterring settlers from reaching the fertile west coast. That changed during the Gold Rush of 1848–1849, when people flocked to California in the hope of finding a fortune. Death Valley earned its name when 12 people out of a party of 30 died trying to find a shortcut to the goldfields in 1849. In 1859 the Comstock Lode gold and silver deposits were found in Nevada, leading to a boom in the population around Virginia City. Mining remained the mainstay of Nevada's economy for years, though many of the settlements are now ghost towns.

One notable group of settlers were the Mormons, who, having fled persecution in the east, in 1847 chose an empty valley near Great Salt Lake as their new home. The settlement grew into Salt Lake City, now capital of Utah. The Great Basin is still one of the most sparsely populated parts of the United States, though the population is growing rapidly. Thanks to Nevada's legalization of gambling in 1931, the tourism and entertainment industries now dominate the economy.

**Bristlecone pine trees are the oldest living organisms on earth, with a lifespan of up to 5,000 years. They are found on the mountaintops of the Great Basin.**

**The low altitude and relentless sun combine to make Death Valley one of the hottest places on earth. The valley floor – a scorching, sun-baked salt flat – is devoid of life.**

The unique landscape of the Great Basin was shaped by elemental forces over millions of years. Like much of western North America, the land here was deformed by the collision between the continental plate of North America and the plate that forms the Pacific seabed. Moving inch by inch, the ocean floor pushed under the continent, causing the land to crumple and rise up. For complex reasons the crust below the Great Basin later began to subside and thin out, a process that is continuing today. As well as sinking, the crust stretched, and this produced massive cracks called faults running from north to south. Huge blocks of crust between the faults slipped down, resulting in the alternating valleys and mountain ranges that characterize the region. Death Valley – the deepest valley of all – is mostly below sea level. Like other valleys in the Great Basin, it has a flat floor where the valley bottom filled with sediment carried by rivers and streams in wetter times.

Most of the world's deserts lie in belts between the tropical and temperate regions, but the Great Basin has a more northerly latitude and a correspondingly cooler climate. The region's low rainfall is a consequence of geography more than latitude. Moisture-bearing winds from the Pacific Ocean are pushed upward by the Sierra Nevada, causing the water vapor they carry to condense and fall on the mountains as snow and rain. Consequently, the air flowing past the mountains and over the Great Basin is much drier, and the region lies in a rain-shadow (*see* pp. 18–20). However, large storms do occasionally cross the mountains and bring downpours to the desert, temporarily filling the streams and sinks. About 180–300 millimeters (7–12 in.) of rain falls on the Great Basin annually, and this is spread across the year more evenly than in the other North American deserts. In winter the rain often falls as snow.

# Mojave/Sonoran

**1. Mojave Desert**
This arid corner of southeastern California lies where the hot Sonoran Desert merges into the cooler and higher Great Basin. The Mojave is distinguished by its unique vegetation, especially the Joshua tree (*Yucca brevifolia*; *see illus.*, pp. 120–121). Like the Great Basin, the landscape is made up of rugged hills and low-lying valleys with salt flats.

**2. Joshua Tree National Monument**
The Joshua tree, a treelike yucca plant, is found only in the Mojave Desert and is especially common in this reserve. The trees are famous for their grotesquely gnarled and twisted shapes.

**3. Palm Springs**
The resort of Palm Springs, California, is an opulent, artificial oasis, popular with movie stars and millionaires. The shadow cast over the town by the massive San Jacinto Mountains brings welcome relief from the sun.

**4. Salton Sea**
This was a desiccated salt flat until 1905, when the Colorado River broke its banks and inundated the area. The lake – about 932 square kilometers (360 sq. mi.) in area – is now sustained by runoff from irrigated farmland, but evaporation has made it as salty as seawater.

**5. Colorado Desert**
The arid region around the Salton Sea is called the Colorado Desert and is part of the Sonoran Desert. Storms from the Pacific can trigger spectacular spring flowerings.

**6. Imperial Valley**
This valley, partly in the southeast corner of California and partly in Mexico, was formerly uninhabited arid desert. Successive irrigation projects (the first completed in 1901) have turned the area into an intensive agricultural region growing sugar beets, cotton, citrus fruits, and alfalfa.

**7. Organ-pipe Cactus National Monument**
This national park in south Arizona was established in 1937 to protect the abundant wildlife in this mountainous part of the Sonoran Desert. It is named for species of the organ-pipe cactus (*Lemaireocereus thurberi* or *L. marginatus*), whose tall branches grow in huge clusters like organ pipes.

**8. Sonoran Desert**
This arid area of North America sprawls through southwest Arizona, southeast California, west Sonora state, Mexico, and Baja California Norte state, Mexico. The desert includes the Colorado Desert, the former salt flat of the Salton Sea, and the intensively irrigated Imperial Valley.

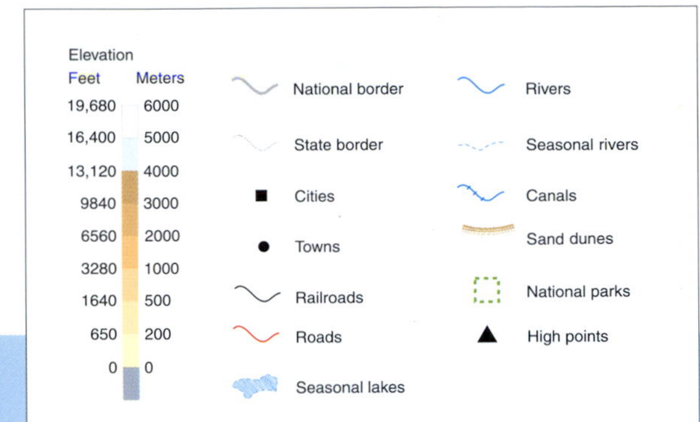

**9. Saguaro National Park**
The towering saguaro cactus (*Carnegiea gigantea*), with its curving side-branches, is perhaps the most distinctive sight in the Sonoran Desert. Saguaros can grow 15 meters (50 ft.) tall and live up to 200 years. Forests of the cacti can be seen in this beautifully scenic park in Arizona.

**10. Los Angeles**
The second-largest city by population in the United States, Los Angeles lies on the edge of one of the country's most arid regions. The city's water supply is brought from the Colorado River by an aqueduct.

**11. Yuma Desert**
The Yuma Desert is a dry and sandy part of the Sonoran Desert that straddles the United States–Mexican border to the south of Salton Sea. The flat land is dotted with bushes of creosote, burr sage, and other desert scrub plants.

**12. Baja California**
This mountainous peninsula in Mexico is sometimes considered a part of the Sonoran Desert, despite its distinct climate and flora. Desert plants include giant cacti, boojum trees, and elephant trees. Volcanoes and frozen lava flows dominate the center of the peninsula. Toward the south is the small Vizcaíno Desert.

**13. Colorado River**
One of North America's major rivers, the Colorado is 2,330 kilometers (1,450 mi.) long and drains a vast, arid area between the Rocky Mountains in Colorado and the Gulf of California in Mexico. Water from the river is diverted by means of aqueducts to supply the urban needs of cities such as Los Angeles, Phoenix, and San Diego.

**14. Kitt Peak Observatory**
The cloudless desert sky and high altitude create perfect conditions for star gazing at this mountaintop observatory.

**15. Montezuma Castle National Monument**
Nestled into a limestone cliff over the valley of the Verde River, this superbly preserved 600-year-old ruined cliff dwelling was built by Sinagua Indians, who are also known as the Western Anasazi. The building has five stories and 20 rooms.

**16. Salt River**
This 322-kilometer-long (200 mi.) river is generally dry in its parts below Phoenix. This is not caused by seasonal aridity, but is the consequence of the series of upstream dams (including the Theodore Roosevelt, Horse Mesa, and Stuart Mountain dams), which create a 97-kilometer (60-mi.) chain of lakes.

## Fact File

- The Mojave Desert covers some 64,750 square kilometers (25,000 sq. mi.) in southern California.
- Summers in the Mojave are hot and windy, but in winter there are frequent frosts. Extreme variations in daily temperatures occur, with cold nights followed by scorchingly hot days.
- The salt flats of the Mojave Desert yield minerals such as borax, potash, and table salt (sodium chloride). There are also deposits of gold, silver, tungsten, and iron in the desert.
- About 200 plant species are native to the Mojave Desert, including the Joshua tree and the Mojave yucca. Other plants include sagebrush, creosote bush, bladder-sage, and catclaw (a type of acacia whose stems are covered with curved claws). Despite the arid climate, there is enough vegetation to support cattle ranches in the northern part of the Mojave.
- The Joshua tree was given its name by Mormons, who saw the curving branches as the arms of Joshua beckoning them toward the promised land.
- The Sonoran Desert's name probably derives from the word *sonota*, which means "place of plants" in the language of the Tohono O'Odham (Papago) Native Americans.
- Trees are more abundant in the Sonoran Desert than the Mojave, especially in hilly or mountainous regions. Among the most common species are paloverdes, desert ironwoods, velvet mesquite, ocotillos, and catclaws. The lowlands are dominated by shrubs such as desert saltbush, wolfberry, and creosote bush. Fir and juniper can also be found on the mountains, and trees such as willow, ash, and walnut thrive along streams.
- The saguaro cactus grows to only 2 centimeters (0.8 in.) in height in its first 10 years of life and may take up to 200 years to reach a maximum height of 15 meters (50 ft.). It reaches sexual maturity at 50–75 years of age, when it bears white flowers (the Arizona state flower) and red fruits.
- Animal life in the Sonoran Desert is abundant and diverse. Among the many reptiles are venomous lizards called gila monsters, desert tortoises, chuckwallas, geckos, and numerous species of rattlesnakes. The mammals include pronghorn antelope, peccaries, coyotes, bobcats, mule deer, cougars, bighorn sheep, pocket gophers, pack rats, kangaroo rats, and squirrels. Desert birds include quail, cactus wrens, elf owls, gila woodpeckers, and white-winged doves.

## The Green Desert

The Sonoran Desert is the hottest of the North American deserts, yet it is also one of the greenest. It stretches across the southwest United States into northern Mexico and covers the southern half of Arizona, southeastern California, and much of the Mexican state of Sonora. Some authorities include most of the Mexican state of Baja California Norte and the islands of the Gulf of California as well. Altogether, the Sonoran Desert occupies about 260,000 square kilometers (100,000 sq. mi.).

Apart from where it meets the Pacific Ocean, the Sonoran Desert does not have any definite boundaries. In the north it merges into the Mojave Desert, an arid zone of valleys and mountains like the Great Basin but distinguished by its unusual vegetation. In the east the Sonoran rises to meet the higher Chihuahua Desert of Mexico, while in the south it peters out as the climate gives way to wetter, more tropical conditions. Subdivisions of the Sonoran are sometimes given their own names. Different parts of the dry, sandy area near the Salton Sea, for instance, are known as the Colorado Desert and the Yuma Desert.

Compared to other deserts, the Sonoran is fairly lush and its winters are mild. Most rain tends to fall at two times of the year. Storms from the Pacific sometimes soak the parched ground in winter, triggering the synchronized emergence of millions of flowering plants in spring. Called ephemerals, these plants bloom only in occasional years and can transform the landscape into a riot of color. A well-developed summer monsoon from July to mid-September brings wet tropical conditions and sometimes violent thunderstorms. This rain nourishes more than 2,000 desert plant species, including many cacti, among them the tall saguaro cacti and organ-pipe cacti that give the Sonoran Desert its unique look.

Warm winters attract tourists from colder parts of North America to the desert resorts of Palm Springs in California and Tucson in Arizona. The dry air and constant sunshine have

made both towns popular with the health-conscious and with elderly people seeking a comfortable place to retire. Palm Springs is a magnet to Los Angeles' wealthier residents. The Betty Ford Center, where the rich and famous go to rid themselves of drug and alcohol problems, is in nearby Rancho Mirage.

Ancient cultures of the Sonoran Desert include the Sinagua (or Western Anasazi) and the Hohokam. The Sinagua lived between Flagstaff and Phoenix from A.D. 500 to 1300. Like the Anasazi, they built multistoried cliff dwellings, some of which survive. The Hohokam lived in southern Arizona between 200 B.C. and A.D. 1450. They constructed an ingenious network of canals to irrigate crops, and made highly decorative pottery as well as feather and seashell ornaments. Recent desert inhabitants include the Apache Indians, who fought a long-running war against U.S. forces in the 19th century, led by the warrior Geronimo (1829–1909).

***Above left*** **A stand of Saguaro cacti. These giant cacti develop curving side-branches as they mature, sometimes resulting in a distinctive candelabra shape.**

***Left*** **Joshua trees are found only in the Mojave Desert, where they have to contend with rocky soil and extremely hot, dry conditions.**

***Right*** **This spectacular cliff dwelling at Montezuma Castle National Monument in Arizona was built by the Sinagua Indians. The Sinagua disappeared around A.D. 1300 for unknown reasons.**

# The Sechura

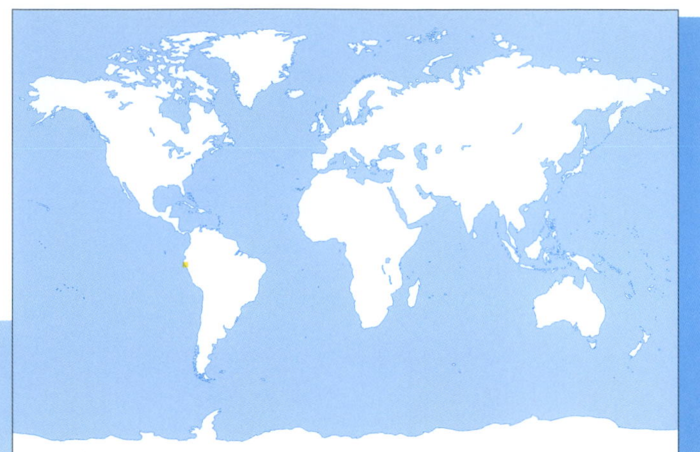

**1. Humboldt current**
This cold current, which flows toward the equator off the Chilean and Peruvian coast, helps maintain the aridity of the western coast of South America, including the Sechura Desert. Air currents flowing onto the land from the sea are cool and carry little moisture, helping to maintain the region's extreme aridity and causing thick fogs where they meet the warmth of the desert sands.

**2. "La Niña"**
In 1997–1998 successive months of torrential rain caused by the El Niño weather phenomenon created a 300-kilometer-long (186 mi.) lake in the Sechura Desert, dubbed locally "Lake La Niña." Owing to its size, the lake is likely to remain a feature of the Sechura for many years and has created a new wildlife habitat.

**3. Piura**
To the north of the Sechura Desert is Piura, the oldest Spanish city in Peru, founded by the explorer Francisco Pizarro in 1532. The city – capital of the Piura department of Peru – is a virtual oasis in the surrounding desert and semidesert landscape, and the region's cotton, rice, citrus-fruit, and sugarcane plantations are maintained by extensive irrigation along the Piura River valley. Piura was severely damaged by an earthquake in 1912.

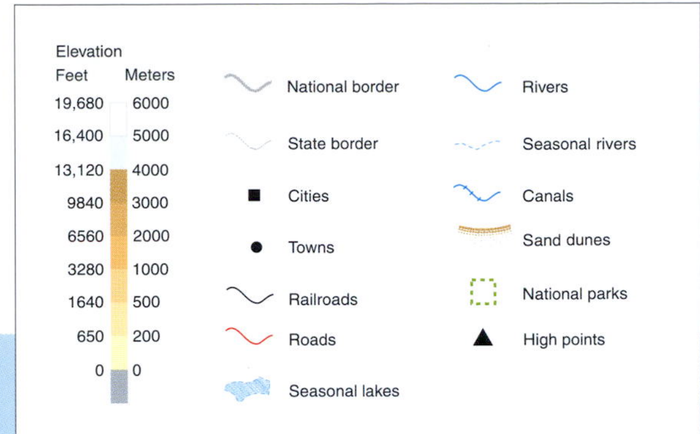


**4. Piura River**
The most important river of the Sechura region, the Piura flows only during the summer, which is the rainy season.

**5. Lowest point**
The Sechura Desert includes the lowest point in Peru, the Bayóvar Depression, which is 37 meters (120 ft.) below sea level.

**6. Ancient rivers**
A feature of Peru's arid coastal region is the dry ravines that mark the flow of ancient rivers. The channel of the Casajal in the Sechura is marked by bushes that grow along its former course. Blowing sand from the surrounding flat terrain, moving inland with the prevailing wind, accumulates around the bushes and builds sand formations known as coppice dunes, which form elevated ridges, marking the earlier course of this river.

**7. Ancient peoples**
The arid coastal region of northern Peru was the home of two important ancient cultures – the Vicus (formerly known as the Sechura) culture, which flourished from about 500 B.C. to A.D. 500, and the Mochica culture (200 B.C.– A.D. 600). The peoples of these cultures were farmers as well as warriors, carrying out extensive irrigation of the region.

**8. Lambayeque**
In 1989 archaeologists discovered the tomb of a wealthy Mochica prince beneath a pyramid near the town of Lambayeque on the southern edge of the Sechura Desert.

**9. Phosphates**
The Sechura Desert has some of the world's richest deposits of phosphorite, or phosphate rock. Phosphates, derived from the shells of marine invertebrates or the bones and excrement of vertebrates, are widely used to make fertilizers.

**The dry landscape of the Sechura Desert supports almost no vegetation.**

# Fact File

- The Sechura Desert forms part of Peru's narrow coastal plain (the Peruvian Costa) between the Pacific Ocean and the Andes mountain ranges – one of the world's driest regions.
- The Sechura Desert is relatively small: its north–south extent is about 105 kilometers (65 mi.); its east–west extent is about 65 kilometers (40 mi.).
- The Sechura is unusual among the world's deserts in its proximity to the equator, which lies only about 645 kilometers (400 mi.) to the north. The desert's tropical location – on the same latitude as parts of the Amazon rain forest to the east – is not matched by a tropical climate.
- Like the Atacama in Chile to the south, the Sechura is a cold-water coastal desert, its aridity maintained by the cold Humboldt current, which runs northward offshore. Other contributory factors are the Andes Mountains, which block the flow of moisture-bearing winds from the Amazon rain forest, and the dry South Pacific high-pressure weather system.
- Average temperatures in the Sechura are cool: 19°C (66°F) in the winter and 22°C (72°F) in summer.
- The Sechura features extensive areas of barchan dunes – crescent-shaped dunes formed where the wind blows almost exclusively from one direction.
- As in the Namib in South America, the little vegetation that survives in the Sechura is supported by moisture from a dense coastal fog – known in Peru as the *garúa*.
- Peru's climate is regularly disrupted by the climatic phenomenon known as the El Niño – part of a wider Pan-Pacific phenomenon that reverses the region's sea and atmospheric conditions. Approximately once every decade the waters off the Peruvian coast become warm, and heavy rains fall on the country's desert regions. By contrast, Peru's normally rainy southern highlands suffer drought. El Niño can have catastrophic effects on Peru's ecology and economy.
- As in all of Peru's arid coastal region, both plant and animal life is rare. Coastal fogs support a mixture of grasses, known locally as *lomas*, and parts of the desert are covered by epiphytes or thickets of algarroba (mesquite) and sapote. Only a few animals, such as the Sechura Desert coral snake, are able to survive the desert's extreme aridity. The coastal waters, by contrast, have a rich variety of bird, marine mammal, and fish life.

# The Atacama

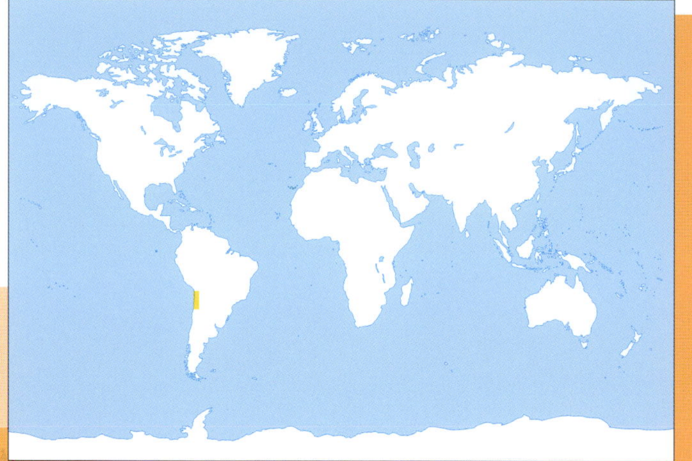

**1. Humboldt current**
This cold current, which flows northward off the Chilean coast, helps maintain the Atacama Desert. Air currents flowing onto the land from the sea are also cool and carry little moisture, helping to maintain the region's extreme aridity.

**2. Cordillera Domeyko**
This mountain range marks the eastern edge of the Atacama Desert. The arid land around the Cordillera Domeyko consists largely of pebble-covered alluvial fans sloping down the foothills.

**3. Cordillera de la Costa**
These low, coastal mountains form the western edge of the Atacama Desert. They reach 1,500 meters (5,000 ft.) in height and in many places end abruptly in towering sea cliffs.

**4. Antofagasta**
The largest city in the Atacama Desert is Antofagasta, a seaport established in 1870 to export minerals from nitrate mines. The nitrate industry has since collapsed, but Antofagasta remains a major port for Chile's copper exports. Heavy rain falls here only two to four times a century.

**5. Copiapó**
This silver- and copper-processing town marks the southern limit of the Atacama.

**6. Atacama Salt Flat**
The waters of this once vast lake have now mostly evaporated, leaving behind a parched salt flat, known in Spanish as the Salar de Atacama. Laguna Chaxa in the middle of the salt flat is a shallow pool of brackish water inhabited by several species of flamingos (*see illus. opposite*).

**7. Valle de la Luna**
In the "Valley of the Moon," in the Atacama Salt Flat, is a bizarre landscape of eroded rock formations that change color as the sun rises and sets.

**8. Chuquicamata**
Chuquicamata in Chile is the world's largest opencast copper mine. The copper ore is quarried from a gigantic, artificial pit some 4 kilometers (2.5 mi.) long and 630 meters (2,100 ft.) deep.

**9. Llullaillaco**
At 6,723 meters (22,057 ft.) high, this volcano in the Andes, overlooking the Atacama Desert, is one of Chile's tallest peaks. Other peaks bordering the Atacama are Cerro Rincón (5,594 m.; 18,353 ft.), Mount Sapaleri (5,652 m.; 18,545 ft.), and Mount Aucanquilcha (5,118 m.; 16,792 ft.)

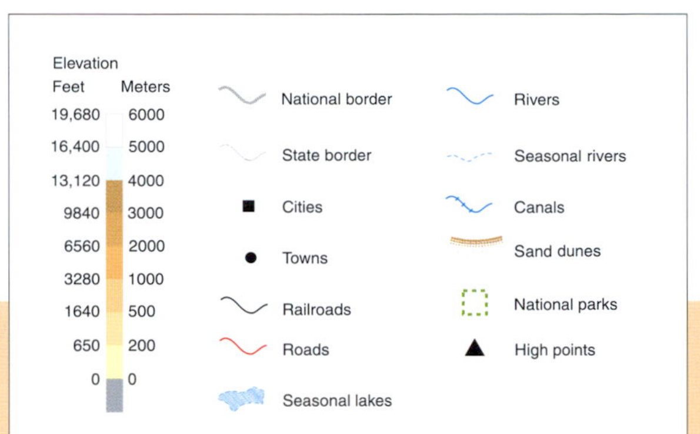


**10. San Pedro de Atacama**
This scenic oasis town just to the north of Atacama Salt Flat is a popular destination for tourists. One of the town's sights is the Museo Gustavo Le Paige, a museum crammed with pre-Columbian artifacts that have been preserved by the region's cool, dry climate, including 9,000-year-old mummies.

**11. El Tatio Geysers**
Close to San Pedro de Atacama are the El Tatio Geysers – spouts of boiling water that are all the more spectacular for their dramatic setting – at 4,300 meters (14,000 ft.) altitude surrounded by snowcapped peaks.

**12. Loa River**
The winding Loa River is the only perennial river in the Atacama Desert. Flowing from the Andes to the Pacific, it is 445 kilometers (275 mi.) long. Water from the Loa maintains the oasis of Calama (now a service city for the mining industry), where Native Indians once irrigated the land to grow crops.

**13. Uyuni Salt Flat**
Covering more than 10,000 square kilometers (3,860 sq. mi.), this is one of the world's largest salt flats. Known in Spanish as the Salar de Uyuni, the salt flat lies in a depression between the western and eastern Andes in the Bolivian Altiplano.

**14. *Nomad***
In 1997 scientists in Carnegie Mellon University in Pittsburg, Pennsylvnia, tested the interplanetary robot *Nomad* on the moonlike surface of the Atacama Desert, directing the wheeled rover over a 200-kilometer (124-mi.) trek from more than 8,600 kilometers (5,000 mi.) away.

**15. Tocopilla**
This old seaport to the north of Antofagasta specializes in the processing of copper ore and iodine, minerals found in the Atacama Desert.

## Fact File

- The Atacama Desert is a strip of land some 1,100 kilometers (700 mi.) long that runs parallel to the coast of northern Chile, bounded in the west by a range of coastal mountains and in the east by the foothills of the Andes.
- The desert floor has an average elevation of about 610 meters (2,000 ft.) above sea level and is made of up of gravel-covered regions, salt flats, and some sand dunes.
- Often described as the driest place on earth, the Atacama receives less than 13 millimeters (0.5 in.) of rain each year. Some parts have received no recorded rainfall, and inhabitants of some of the towns have never seen rain in their lifetimes.
- Temperatures in the Atacama Desert are relatively low because of the cooling effect of the Humboldt current. The average summer temperature in Antofagasta is only 18ºC (65ºF).
- The Humboldt current produces a heavy mist over the Atacama's coast in the mornings. Dew from this mist sustains sparse vegetation and animals such as insects and lizards.
- For many years the Atacama was one of the world's major sources of nitrates, used in the production of fertilizers and explosives. In 1879–1884 Chile and Bolivia (in whose territory Atacama then lay) went to war over mining rights in the region. At the end of what became known as the Pacific War Bolivia ceded its only coastal province to Chile.

**Flamingos are among the few animals able to survive in the Atacama's salty lakes.**

## A Waterless World

The Atacama Desert's reputation as the driest place on the earth is well deserved. So little rain falls on this strip of northern Chile that some parts of the desert experienced a 400-year drought lasting until 1971. The extremely arid climate is caused by a combination of several factors. First, like many of the world's major deserts, the Atacama lies at a subtropical latitude where the climate is dominated by high pressure and the downward movement of warm, dry air masses. Second, the cold Humboldt current along the coast of South America reduces evaporation from the ocean, so air moving over the water picks up little moisture; and third, the Andes block the movement of humid air from the Amazon, casting a desiccating rain-shadow over the region.

Plant life in the central desert is almost nonexistent and few animals survive there. However, the nutrient-rich waters of the Humboldt current support a diverse coastal fauna, including otters, sea lions, and many species of seabirds, such as Humboldt penguins, pelicans, and petrels. Inland, where the land rises in the eastern Atacama, cacti appear, as well as lizards, insects, chinchillas, and rabbitlike rodents called vizcachas. Vicuñas and guanacos – wild relatives of llamas – live on the sparse grass and scrub of the Chilean highlands, as do herds of ostrichlike flightless birds called rheas.

Much of the desert consists of gravel-covered plains, but there are also rolling sand dunes reminiscent of the Sahara, salt flats like those of the Great Basin, and canyons cut by rivers that once flowed from the Andes. Where the land rises in the east, the scenery is dominated by cone-shaped volcanic peaks that create dramatic skylines.

# PEOPLE OF THE DESERT

Despite the harsh climate, people have lived in the Atacama Desert or on its fringes for at least 10,000 years. Among the earliest cultures were the Las Conchas people, who lived on the coast near Antofagasta around 7500 B.C. and survived on a diet of fish, shellfish, and sea-lion meat. The Las Conchas buried their dead with mortars containing hallucinogenic seeds, which suggests that drug-oriented rituals played an important role in their religion.

Around 5000 B.C. the Chinchorro people of the Atacama coast began a practice that was to continue in South America until the Spanish conquest: mummification of the dead. Rather than preserving bodies whole, the Chinchorro first dismembered the corpses and defleshed the bones. The skeletons were then reassembled and reinforced with sticks, and internal organs were replaced with bundles of reeds or other materials. The skin was put back, with seal skin to patch any gaps, and the entire body was painted with ash paste and a coat of shiny black manganese. Decorative clay masks and real-hair wigs provided the finishing touches.

Agriculture reached the Atacama between 5000 and 2500 B.C. In pre-Columbian times the Atacameño Indians grew crops such as maize, beans, potatoes, and cocoa in irrigated land around the desert's oases. The Atacameño had close links with the Andean cultures of Peru and Bolivia and used llama trains to trade fish, guano, maize, and other goods with them.

In the 16th century the Spanish conquest transformed South American cultures. Today's desert inhabitants are largely of Spanish descent and depend for their livelihoods mainly on the region's mining and fishing industries.

***Above*** **Wind-eroded rock formations and salt deposits are a feature of the Valle de la Luna.**

***Left*** **The desiccating air of the Atacama Desert has created vast salt flats.**

***Right*** **This 5,000-year-old mummified woman was found in the northern Atacama Desert in 1983. The clay mask coated with black manganese is typical of Chinchorro mummies. A whalebone placed beside her may be a sign of wealth or status.**

# Patagonia

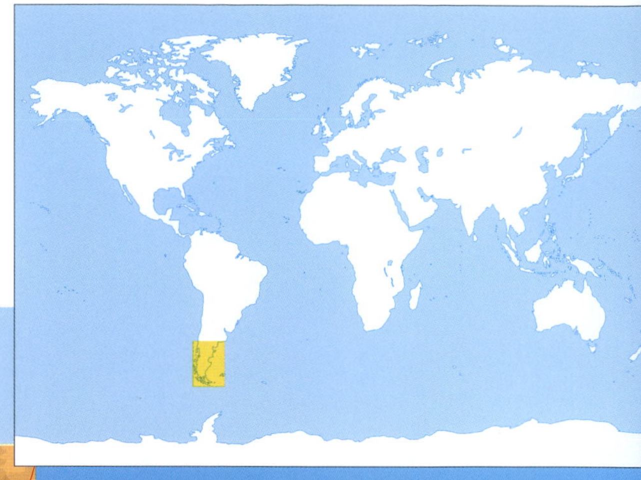

**1. Santa Cruz**
Expanses of dark, solidified lava cover large parts of southern Patagonia. Patchy yellow grasses give the ground a leopard-skin appearance in places. Santa Cruz and neighboring Chubut are home to Argentina's oil industry.

**2. Route 40**
The road snaking through the center of southern Patagonia is Route 40. This largely gravel track winds through a landscape of flat-topped mesas, grassy valleys, and occasional *estancias* (sheep ranches).

**3. Chubut River**
Like most of Patagonia's small rivers, the Chubut flows from the Andes to the Atlantic in a deep, wide valley bordered by high cliffs. This canyon was carved out of the high ground when the climate was wetter.

**4. Neuquén**
Numerous dinosaur fossils have been discovered near this desert town, including *Giganotosaurus*, a predator bigger than *Tyrannosaurus rex*.

**5. Río Negro**
The Río Negro region is the only part of Patagonia with extensive agriculture, thanks to a dam on the Negro River that provides water for irrigation. Crops include apples, pears, grapes, peaches, almonds, and alfalfa.

**6. Patagonian Ice Sheet**
The ice sheet covering the southern Andes is the world's third-largest ice field. The land around it shows many signs of past glaciations, such as fjords, glacial lakes, and U-shaped valleys.

**7. Lake Viedma**
This is one of a chain of lakes between the Andes and the Patagonian tableland. It was scoured out of the ground by glacial erosion and dammed by rubble dropped by glaciers.

**8. Parque Nacional los Glacieres**
A popular tourist destination, this park incorporates Lake Viedma, Lake Argentino, and glaciers of the Patagonian ice sheet. Wooden catwalks allow visitors to walk close to the spectacular Moreno Glacier to watch icebergs break off it and crash into Lake Argentino.

**9. Parque Nacional Perito Morena**
This national park incorporates more than 115,000 hectares of mountains and lakes. Southern Patagonia's tallest mountain, Monte San Lorenzo, dominates the skyline.

**10. Strait of Magellan**
Many ships have been wrecked in this passage between the Atlantic and Pacific oceans due to the ferocious westerly winds.

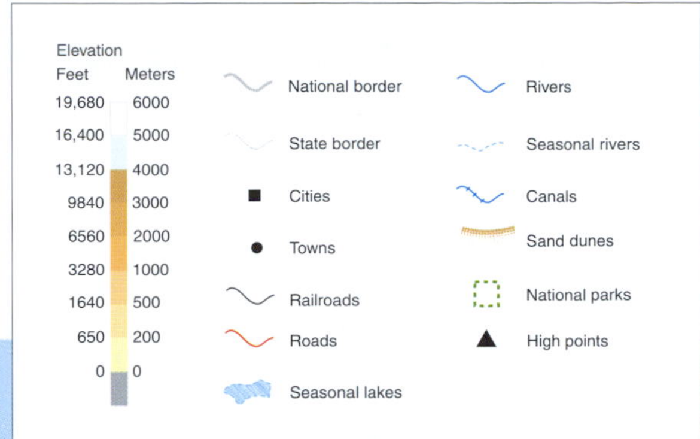


**11. Tierra del Fuego**
The Tierra del Fuego islands were discovered in 1520 by the Portuguese explorer Ferdinand Magellan (*c.*1480–1521), during his famous trip around the world. Native Indians lived here as hunter-gatherers until the late 19th century.

**12. Deseado River**
The Deseado River in central Patagonia is barely a stream and sometimes dries up completely. Windblown sand has buried some parts of it, leaving almost no sign of the river's existence.

**13. Atlantic coast**
Much of Patagonia's long Atlantic coast consists of a narrow plain backed by tall cliffs. The sheltered beaches here provide breeding grounds for numerous penguins and marine mammals.

**14. Valdés Peninsula**
Colonies of Magellan penguins, elephant seals, and sea lions flourish on the remote Valdés Peninsula in central Patagonia. In autumn the sea-lion colonies are attacked by killer whales.

**15. Comodoro Rivadavia**
Comodoro Rivadavia in Chubut province is the largest city in Patagonia, with a population of more than 100,000. The local economy depends almost entirely on nearby oil fields.

**16. Rawson**
Rawson and other towns in the Chubut Valley were founded by Welsh settlers in the 19th century. For many years Welsh was the main language spoken in this part of Patagonia.

## Fact File

- Patagonia is a vast, windswept, treeless plateau. Its arid climate is caused by the Andes, which push up the prevailing westerly winds and so cast a rain-shadow over the area.
- With an area of about 670,000 square kilometers (260,000 sq. mi.), this is the largest arid zone in the Americas. It stretches from the Andes to the Atlantic, occupying most of southern Argentina between the Colorado and Coig rivers.
- Because of its high latitude, Patagonia is a relatively cold desert. The average annual temperature is about 16ºC (61ºF) in the north and 8ºC (46ºF) in the south. Subzero conditions are common in winter, when heavy snow falls in the south. Summers in the north can be blisteringly hot, however.
- Desert wildlife includes pumas, maras (Patagonian hares), armadillos, skunks, and rare herds of guanacos and rheas. The sparse vegetation consists mainly of low shrubs and grasses.
- The name Patagonia may come from Patagon, a dog-headed monster in Spanish legend, or it may derive from the Spanish for "big feet." In either case *Patagones* was the name given by early Spanish explorers to the native peoples of the region.

**Lake Argentino in Santa Cruz province lies between the Andes Mountains and the bleak, treeless plains of Patagonia.**

# THE DESERT IN HISTORY

*People have lived in deserts for thousands of years. Nomadic hunter-gatherers managed to adapt to desert conditions. In Asia rich oasis cities grew up along trade routes, while in North America thriving cliff settlements perched on mesas or nestled against canyon walls.*

Some of the world's oldest and most powerful cities have flourished at desert oases or in close proximity to desert environments, and their names are often rich with historical and legendary associations. The biblical stronghold of Jericho (the Palestinian Arihā), which famously fell to the Hebrew commander Joshua, grew up beside a desert spring and has been inhabited continuously for some 10,000 years. In the fourth millennium B.C., a cluster of cities, including Ur and Erech, developed in arid Mesopotamia, as settlers learned to irrigate even the most unpromising of its lands, and laid the foundations of one of the earliest civilizations.

In Roman times trade between the Mediterranean and East Asia gave rise to great cities in the Syrian Desert, such as Petra and Palmyra. In arid Central Asia Bukhara and Samarqand were among the richest cities of the Persian Empire. Although Samarqand was destroyed by the Macedonian king Alexander the Great in 429 B.C., during the first millennium A.D. the city flourished once again as one of the principal pivots of the great trans-Asian trade route – the Silk, or Golden, Road.

**Reputedly founded by the Hebrew king Solomon, the ancient city of Palmyra (the biblical Tadmur) flourished on an oasis on the northern edge of the Syrian Desert. Under Roman rule the city became an important stopover on East–West trade routes, but was largely destroyed after Emperor Aurelian crushed a local revolt led by Palmyra's queen, Zenobia, in A.D. 272.**

Despite such notable examples, settled life even in the desert margins has always been precarious, and the desert city is in many ways a creation of human ingenuity that flies in the face of nature. The spectacular ruins of Petra and Palmyra are mournful reminders of just how susceptible such towns and cities were – and are – to changing political and economic fortunes. As the lucrative East–West trade faded, Petra, for example, became a desert outpost of the Byzantine Empire, and in its enfeebled state became easy prey, first to Muslim and later to Crusader warriors.

The most successful desert dwellers of history, perhaps, were not those who settled in the desert but those who moved about it, wandering from place to place in search of food, water, or pasture in the light of centuries of experience, handed down as hallowed desert lore. Despite, or even because of, the rigors of their

## AT THE DESERT'S EDGE

Every scrap of fertile soil was precious to the ancient Egyptians, so they built pyramids and buried their dead just beyond the cultivated areas in the necropolises that grew up on the edge of the desert. This almost certainly gave rise to one of the most characteristic features of Egyptian civilization – mummification. The poor could not afford elaborate tombs and were buried in simple pits with a few grave goods. One effect of this burial method was that the surrounding sand drew all of the moisture from the corpses, leaving them desiccated but intact; by contrast, the corpses of the wealthy, buried in elaborate tombs, rapidly decomposed. Observation of this natural mummification encouraged the Egyptians' belief in the importance of physical survival after death and prompted them to develop the techniques of artificial mummification they famously practiced on the pharaohs buried in the rock-hewn tombs of the Valley of the Kings.

existence, many of these nomadic peoples were able to develop rich, complex cultures that, even in the absence of towns and cities to act as memorials to their achievements, have left behind a heritage of great beauty. Nomadic desert-dwellers have often been great warriors, too, and their intervention in the affairs of settled communities has at times dramatically changed the course of human history.

### ANCIENT CIVILIZATIONS

The desert environment has never been merely a barrier to human endeavor, but on occasion has acted as a stimulus to cultural development. Civilizations – here taken to mean those cultures in which sufficient food is produced for substantial numbers of the population to engage in non-subsistence activities such as art, administration, and warfare – often seem to develop not where life is easy, but where the prevailing conditions are challenging enough to elicit a strenuous and creative response. Unpromising environments, apparently in thrall to the desert, have stimulated their inhabitants to new heights of achievement.

One of the world's earliest civilizations appeared in Mesopotamia, the flat and arid land between the Tigris and the Euphrates rivers, in what is now Iraq. Sometime about 4000 B.C. settlers in the area, which they called Sumer, began to develop a complex system of irrigation to maintain their crops, transforming this unpromising environment into what was essentially a giant oasis, often known as the Fertile Crescent.

A communal enterprise on this scale demanded an unprecedented degree of political and social organization, and administration was in the hands of a priestly class, which oversaw the building and maintenance of the irrigation ditches and canals, distributed food among the community, and stored up grain against periods of famine. Social organization in turn inspired other developments, such as the growth of towns and cities, cuneiform ("wedge-shaped") script, numerals, and a rich artistic tradition that included the world's first great narrative poem *The Epic of Gilgamesh* and elaborate, sumptuously decorated sculpture.

A similar process took place just a little later, when the ancient Egyptians organized to exploit the thick layer of silt left every year by the flooding of the Nile – an exotic river much of whose course ran through barren desert – and laid the foundations of what became pharaonic civilization. While the achievements of ancient Egypt were in no small degree "the gift of the Nile," they were also in part a response to the nearby desert. Upper (southern) Egypt had originally been settled by nomadic Neolithic peoples of the Sahara region, while the North African deserts, which began to form only in

**The pyramids at Giza stand on the edge of the Libyan Desert. For the ancient Egyptians the desert was closely associated with death and the afterlife. Anubis, the god of death, was depicted with the head of a jackal, an animal that dwells on desert margins.**

**The ruins of Ur, in modern Iraq, only begin to suggest the extent of one of the greatest cities of Sumer. Founded in the fourth millennium B.C., Ur flourished until the fourth century B.C., when the nearby Euphrates River seems to have changed its course, turning the city's fields to desert.**

the fourth millennium B.C. after the end of the last ice age, gave both protection and cohesion to the young river kingdom.

Through millennia the desert continued to throw a powerful shadow over Egyptian culture. The greatest god was not Hapi, the deity of the Nile flood, but Ra (or Re), the sun god, whose overwhelming presence in the landscape inspired fear as well as gratitude. The destructive aspects of the sun were represented by Ra's daughter, the lion-headed goddess Sekhmet, while the evil god Seth suggested the chaotic, windblasted desert itself. The precariousness of life in Egypt, dependent on an uncertain annual flood and hemmed in by a hostile desert, encouraged both a festive, celebratory attitude to everyday life and a near obsession with death and the afterlife.

The ancient civilizations of Mesopotamia and Egypt are, however, only the best-known of the many cultures that flourished in or on the margins of deserts. Across the globe this uniquely challenging environment has provided homelands for some remarkable peoples. The scarcity of resources has often necessitated close-knit, enterprising communities, in which sharing is a crucial survival strategy. While in many cases desert life promoted a largely peaceful and sequestered lifestyle, in some instances its rigors have encouraged the most vigorous of enterprises – conquest and war. Many desert peoples survive today, although often their way of life has changed radically under the impact of European colonialism and modern technological and economic hegemonies (*see* Chapter 6).

## Peoples of the Sahara

Ancient engravings of elephants, giraffes, and rhinoceroses on rock faces in the Sahara are evidence of the relatively swift climatic changes that transformed the region from fertile savanna into inhospitable desert between the fourth and second millennia B.C. The paintings are strikingly naturalistic and have some affinities with the more famous Paleolithic paintings of France and Spain.

Among the most ancient recorded peoples of the Sahara and its fertile Mediterranean fringes are a Caucasian people today known by their Arabic name, the Berbers (ultimately deriving from the Greek word for "foreigner"). They were already long established in the region by Classical times, featuring in Roman histories as the Numidians. In the seventh century A.D. Arab invaders defeated the Berbers and converted them to Islam. Four centuries later they were subjugated once again, this time by land-hungry bedouins (nomadic Arabs; *see* pp. 136–137), who displaced them in many parts of North Africa. During this period many Berbers adopted settled Arab lifestyles, but others joined nomads of an ethnically similar background who had already taken to living in the arid lands of the interior.

The most numerous and feared of the Berber nomads were the Tuareg, who eventually came to dominate the entire central Sahara. Though converted to Islam, they proudly maintained their own traditions and continued to use their own language and write it in a distinctive script derived from ancient Libyan. Their society was matrilineal, so that property and rank were inherited through the mother, and Tuareg women enjoyed property rights and a personal freedom unusual in most other Islamic societies. Reversing bedouin practice, Tuareg men were veiled in public but women were not.

The Tuareg were pastoralists, herding sheep, goats, and cattle. The camel was vital to their way of life, both as the pack animal that transported the family group from place to place and the swift and hardy mount on which they could conduct raids and make war. Despite the importance of animal husbandry and trade in their lives, the Tuareg were above all warriors, making war and domination central to both their heroic code and their economy.

The Tuareg social structure consisted of nobles, vassals, and slaves. The majority of the slaves were the descendants of captured black oasis-dwellers or individuals purchased from the south; they grew crops for the Tuareg and did menial work. The Tuareg extorted tribute from oases and protection money from desert caravans, and raided and made war on other tribes. Some of these practices were modified as the French colonized North Africa in the 19th century, with salt trading replacing caravan raiding and Tuareg slaves reclassified as serfs. Remarkably, however, the traditional way of life survived largely intact until the mid-20th century.

## Caravans across the desert

Saharan rock paintings provide evidence that journeys across the North African desert were undertaken in ancient times, using a few routes where conditions were not too hostile for horses and wagons. Despite the progressive desiccation of the Sahara, caravans from Numidia (modern Algeria) reached the Niger and brought back ivory and gold. In A.D. 70, after North Africa became part of the Roman Empire, two military expeditions reached Lake Chad and the Niger. Regular travel and trade on any scale began only after the seventh-century Arab conquest of North Africa, when traders followed the trails blazed by Berber nomads who had begun to cross the desert by camel.

The Arabs were inspired as much by mercantile as religious fervor, establishing a thriving trans-Saharan commerce even as they converted much of West Africa to Islam. Goods from fertile North Africa, and later from Europe, were carried by donkey train to Marrakech, Sijilmassa, Ghadāmis, and other towns on the northern edge of the desert. From there, caravans, often comprising thousands of camels, followed one of several routes crisscrossing the desert. There were also routes from Egypt, running west and south from Cairo or south and west from Asyut, and terminating at Lake Chad (*see* Fig. 1).

Of all the commodities brought from the north, the most precious was salt, for which a huge market existed in the largely saltless African interior; at times, when the caravans arrived at southern cities such as Awdaghost, Tombouctou (Timbuktu), Walata, Gao, and Tadmekka, salt and gold were exchanged weight for weight. Gold was so abundant in West Africa that between the eighth and 17th

**Figure 1 From the seventh century trans-Saharan trade routes flourished, connecting the gold-rich civilizations of West Africa with the Arab and Berber peoples to the north.**

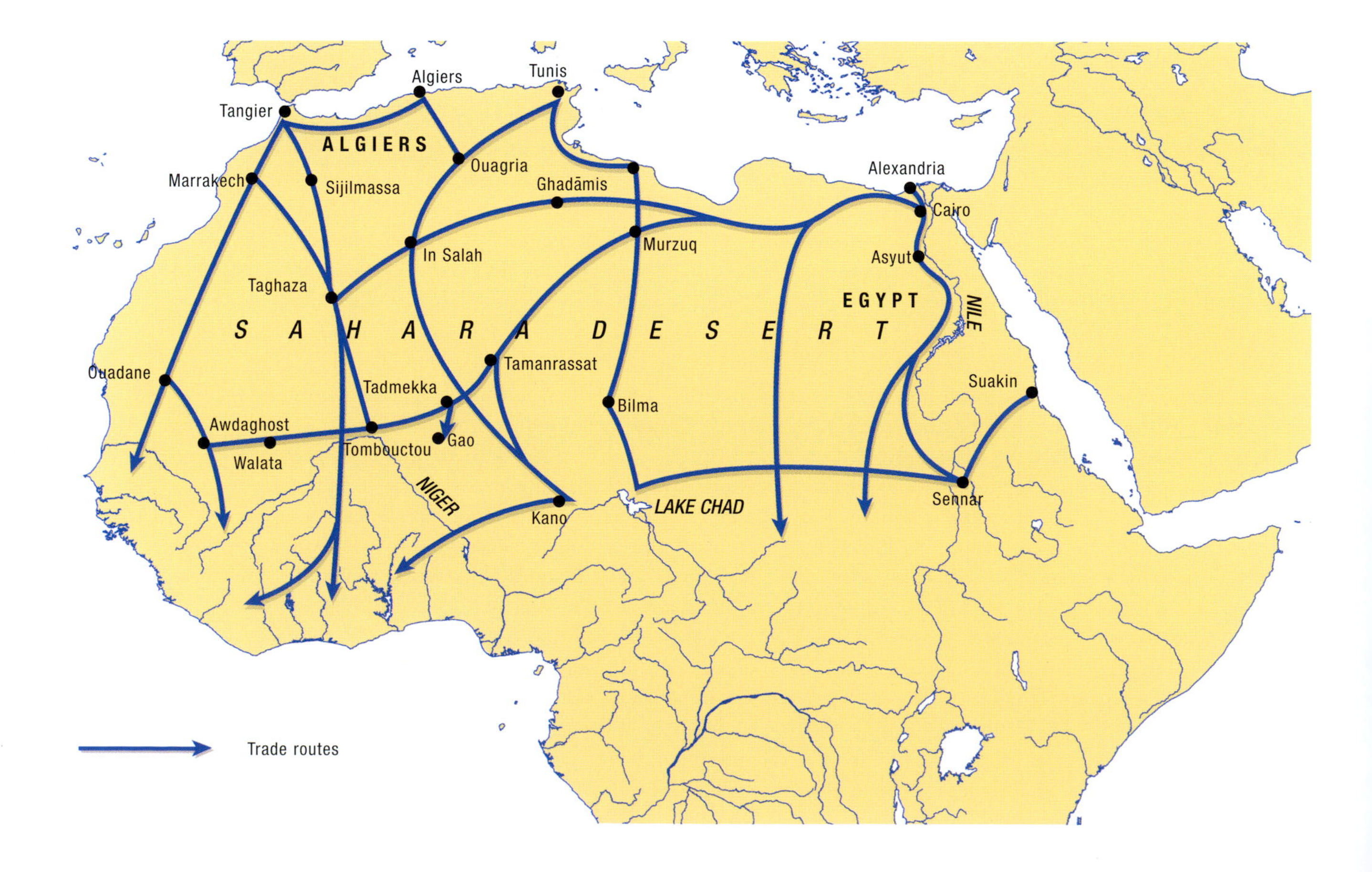

A bedouin camel train crosses the Sahara. Camels remain indispensable in many desert and semidesert regions across the world.

centuries the region supported a succession of wealthy, powerful kingdoms, including Ghana, Mali, and Songhai; in Ghana, it was said, even the dogs' collars were made from gold.

One of the most spectacular trans-Saharan journeys was made in 1324 by Mansa Musa (d. 1332?), king of the Mali Empire, who made a pilgrimage to Mecca, accompanied by a caravan of camels, horses, wives, and slaves. During his stay in Cairo Mansa Musa was so lavish in his gift-giving that the price of gold in the city fell sharply and the local economy took over 20 years to recover. Mansa Musa brought back a group of scholars from his travels who made his wealthy capital, Tombouctou, a celebrated center of Islamic learning.

By the 16th century both the West African kingdoms and the gold trade were in decline, and the political and commercial vacuum was filled by the slave trade. Conditions for the slaves on their journey north were as harsh and deadly in their way as on the ships of the better-known Atlantic trade. French colonial authorities gradually suppressed trans-Saharan slaving in the mid-19th century, and by the time the dauntless traveler René Caillié (1799–1838) reached the fabled "Timbuktu" in 1828, it was a dusty, sleepy provincial town. In some

## DESERT DRESS

Some desert peoples, such as the Aborigines of Australia and the San of southern Africa, have worn few or no clothes. The majority, however, have relied on layers of fabric to protect them against extremes of heat and cold. In North Africa and the Middle East costume for men varied from region to region but consisted essentially of a long, shiftlike cotton garment, worn under an open coat or hooded cloak. Such loose-fitting clothes allowed air to circulate around the wearer's body and slowed down dehydration.

The Tuareg also often wore trousers and garments dyed blue. On their heads Tuareg men wore a turban and face veil, usually made from a single piece of material. Bedouin men covered their heads and the backs of their necks with a large kerchief, secured with bands round the temples. Women's costumes followed the same principles but, on occasions, were bejeweled and decorated with dyed, woven, or embroidered designs.

## HERMITS IN THE DESERT

The desert and the rigors of the life associated with it had a powerful imaginative pull for early Christians. Both Jesus and John the Baptist spent prolonged periods in the "wilderness," struggling with worldly temptations and striving for spiritual perfection, and their example was emulated by hermits, monks, and other Christian ascetics, who sought in the empty barren landscape an ideal of sexual and sensual renunciation that, they believed, brought them closer to God. In the fourth century A.D. the desert fringes of the Nile Delta in Egypt, especially at Nitria, became home to many ordinary pious men and women who built simple cells that were often only a few hours' walk from their native villages. By the end of the century some 5,000 monks were said to live at Nitria alone, with thousands more scattered along the rugged desert fringes of the Nile and the Red Sea. A more extreme renunciation was practiced by the few monks and hermits who lived at Scetis, a day and a night's further journey into the desert from Nitria. These were the so-called Desert Fathers, whose heroic sufferings and spiritual achievements inspired profound awe among later generations of Christians. The struggles of the Desert Fathers were concerned less with sexual temptation than with food. In some cases ascetics were driven half mad with hunger, roaming the desert like wild animals in search of a few roots or herbs. An individual's mastery of this unholy state – known as *adiaphoria* – was a sign of his triumphant humanity. One Egyptian monk scrawled on the wall of his cell as a warning to himself the fatal cry of Esau by which he lost his birthright: "Jacob, my brother, [give me] lentils." It was the desert ascetics' sexual temptations, however, that most exercised the imagination of medieval and Renaissance Christians. The experiences of Saint Anthony of Egypt (*c.*250–355) and Saint Jerome (*c.*347–419 or 420), who spent two years in the Syrian Desert in imitation of the Egyptian Desert Fathers, were of particular fascination and were often portrayed in art. "O how often ... in that lonely waste," Jerome wrote, "scorched by the burning sun ... did I fancy myself surrounded by the pleasures of Rome ... [in] this prison house, where my only companions were scorpions and wild beasts, I often found myself surrounded by bands of dancing girls."

parts of the Sahara, notably Mauritania and Sudan, slavery remains a feature of traditional societies, even today.

### Desert Peoples of the Middle East

The bedouin are the nomads of the Middle Eastern deserts. Arabic-speaking camel herders, they originated in the Arabian Peninsula, eventually moving out into the adjacent desert areas that are now parts of Iraq, Syria, Jordan, and Israel. Converted to Islam, they took part in the Arab invasions launched from the seventh century onward, moving in force into North Africa four centuries later. The main clan to make this long migration, the Beni Hilal, are still much celebrated in bedouin heroic poetry.

Bedouin families traveled in clan groups for support in the desert. The worst punishment an individual could suffer was to be outlawed from his clan group, which virtually condemned him to death. Like the Berber Tuareg, the bedouin were fierce fighters as groups raided caravans and hostile camps. Each clan had its own territory and a strict code of honor meant that disputes between clans frequently ended up in long-running feuds. On the other hand, the bedouin acknowledged their common struggle against the desert by observing a strict law of hospitality: no man could turn away a traveler from his tent.

The bedouin social system was strongly patriarchal, and women were kept from the eyes of strangers or veiled in their presence. The black woolen tents of the bedouin were partitioned into quarters for men, which served as the public area, and the women's quarters, where cooking, weaving, and household tasks were done. The tent itself was made of goat or camel hair, woven into strips and sewn together. Long and low to withstand desert winds, it was supported on poles and arranged so that one or more sides could be rolled up to give the occupants the benefit of any breeze. Cushions, camel saddles, and rugs made the tent a comfortable home, yet one that could be rapidly dismantled when it became necessary to move on.

There was little tillable soil, so crop-growing was rare. Bread was therefore considered a great luxury, its place as a staple food taken by dates from the date-palm. Crushed date stones were formed into cakes and fed to camels. So vital was the date that it was one part – water being the other – of what the bedouin treasured as "the two black ones." Another essential staple was camel's milk. In emergencies, such as if they ran out of water on a journey, the bedouin could use camels in a less likely way. They would force a camel to vomit up the water contained in its stomach to drink, or kill an old camel for the same purpose. If the camel had drunk within a couple of days, the water was said to be tolerable.

The bedouin themselves distinguished between camel herders, the true *badawiyin*, or "desert dwellers," and those who kept flocks of sheep and goats on the edge of the desert, where water was more plentiful (sheep need to drink every day). The camel herders relied on the endurance of their mounts, which also provided them with their staple food – camel's milk. The famous Arab horses, though highly prized, were less useful possessions than status symbols. In times of drought, what water the bedouin had was often reserved for horses and cattle.

Though dynasties and empires succeeded one another across the Middle East, the bedouin remained largely a law unto themselves. Despite their harsh lives, they became the most romanticized of desert peoples – not least by their settled fellow Arabs, for whom they symbolized a mythical past of heroic freedom. Later than most nomad peoples the bedouin continued to take part in wider historical events. During World War I (1914–1918) the Allied-backed Arab Revolt against the Turks was proclaimed by Husayn ibn 'Ali, sharif of Mecca, and bedouin feats of arms contributed to campaigns that led to the fall of Jerusalem and Damascus.

Subsequently Husayn became king of Hejaz (southern Arabia) and his sons Faisal (1885–1933) and Abdullah (1882–1951) founded dynasties in Iraq and Transjordan (present-day Jordan), respectively. The dynasty survives in Jordan, where the loyalty of bedouin troops has proved decisive in several crises. There were further changes in the bedouin heartlands during the 1920s when Ibn Sa'ūd (c.1880–1953), ruler of Nejd (central Arabia), drove Husayn ibn 'Ali from the peninsula and ultimately, in 1932, founded the new state of Saudi Arabia, with himself as king.

## Ships of the desert

Without the camel, human beings would have found it almost impossible to survive and travel long distances in many desert regions of North Africa and Asia. The two-humped Bactrian camel (*Camelus bactrianus*), with its shaggy coat and heavy build, is well adapted to the harsh conditions that prevail in Central Asia. The single-humped dromedary (*Camelus dromedarius*) has played an even more vital role in Arabia, the Middle East, and North Africa, thanks to a unique combination of features. Though valued for its meat, milk, wool, and hides, it is above all a traveler – giving it its name the "ship of the desert" – whether ridden or used as a pack animal.

The camel is a superb example of desert adaption. Long legs lift its head and body away from the hot desert surface into relatively cool air, while soft, spreading footpads pass lightly across the burning sand. The camel's thick eyelashes, furred ears, and ability to close its nostrils offer protection against abrasive windblown sand. The camel can live off coarse or thorny plants, and its fat-filled hump enables it to survive for long periods without eating. It can also go for days without drinking, thanks to body mechanisms that minimize fluid loss, and can shed 25 percent of its body weight by dehydration with no ill effects. Coupled with this iron constitution are strength and stamina – the typical camel can carry about 250 kilograms (550 lb.) and is able to maintain a speed of about 16 kilometers an hour (10 mph) over a long day.

Dromedaries seem to have been domesticated in central or southern Arabia during the second millennium B.C. or earlier. They spread only slowly across the Middle East and into Egypt. In the first century B.C. Julius Caesar saw them in North Africa, and the Romans were using them regularly there by the second century A.D. Their importance became crucial during the Arab conquest of North Africa in the seventh century, making it possible for the newcomers to penetrate deep into the Sahara, carry the message of Islam to the south, and set up trans-Saharan trade routes.

Over the next few centuries camels became the basis of the nomadic way of life in the North African and Arabian deserts. Among desert peoples ownership of camels was the most visible manifestation of wealth, and calculations of value were often made in terms of camels rather than money. The tribes of the western Sahara minimized feuding after a fight or accident by having an agreed tariff of payments as compensation for injuries inflicted. The recompense was expressed in camels, sometimes supplemented by guns.

Among the bedouin camels were, and continue to be, an important element in marriage arrangements. The bridegroom presents his father-in-law with an agreed number of camels, and the bride is carried to her wedding on a camel. On ceremonial occasions camels carried saddles of wood and leather and were decorated with brightly colored woven and embroidered headdresses, neckbands, saddlebags, and cruppers.

**The San peoples of southern Africa produced some of the world's finest prehistoric art. Their rock paintings, found throughout the region and dating from various periods from about 4000 B.C. to the 19th century, are often highly naturalistic depictions of game but are sometimes records of the hallucinogenic imagery experienced by San shamans. This painting of a giraffe is from the Kalahari Desert, in Botswana, southern Africa.**

## CLICK LANGUAGES

The nomadic San of the Kalahari speak a number of different languages, all of which belong to the Khoisan (Khoikhoi–San) group of southern Africa. This small group is notable for its employment of complex systems of clicks, in addition to consonants and vowels. The clicks are produced by moving the tongue or lips without using any air from the lungs; non-African peoples occasionally employ sounds of this kind, for example by blowing kisses or signifying disapproval with the noise usually written as "tut-tut" or "tsk-tsk." In their current written form, the different clicks are represented by the symbols !, //, /, and ≠, and when combined with other sounds, generate large vocabularies. Well-known San language groups include the largest, the !Kung, as well as the G/wi, the Hai//om, and the Nharo.

### Desert Peoples of Southern Africa

The main inhabitants of the Kalahari Desert were the hunter-gatherers called the San or Basarawa. The San have lived in southern Africa for some 20,000 years and occupied much of it until first Bantu peoples and later Dutch colonists drove them from the better lands and confined them to the desert. There the San evolved a way of life that was ingeniously adapted to the Kalahari.

The basic unit of San society was a band of some 20 to 50 people, made up of a number of families that were often interrelated. The bands were part of groups so widely scattered that they spoke different languages and were often unaware of one another's existence. However, the San did have a sense of themselves as the "red people," distinct from both black Bantu-speaking peoples and white colonists. Having no pack animals, they traveled on foot with few possessions. They seldom stopped in one place for more than a few weeks before the ever-present threats of thirst and hunger forced them to move on.

Whenever the band stopped, members built shelters of branches covered with bark or grasses. Hunting was the task of the males, who tracked antelope and wildebeest as well as smaller creatures. Their bows and arrows were not very powerful, so they had to use tracking skills and stealth to get close to their quarry; even then it was not the arrows that killed, but a poison, which was taken from the insides of beetle grubs and smeared onto the weapons' tips. Hunting provided food, sinews for bow strings, and skins for clothing and bags.

The food gathered and prepared by the females was varied and more important than meat in the San diet because of its moisture content; at one time or another they might feast on gathered berries, nuts, roots, bulbs, edible grubs, or birds' eggs. Ostrich eggs were particularly useful, providing nutrition, a receptacle used on occasion to store water, and material from which women could make beads to decorate their hair.

The San way of life has been much admired. There were no leaders, and pooling of resources and gift-giving by the fortunate created a harmonious, egalitarian culture. Food-getting was swift and efficient, leaving plenty of leisure time for dancing, singing, and storytelling. Myths described a creator who was omnipotent and invisible, a lesser being

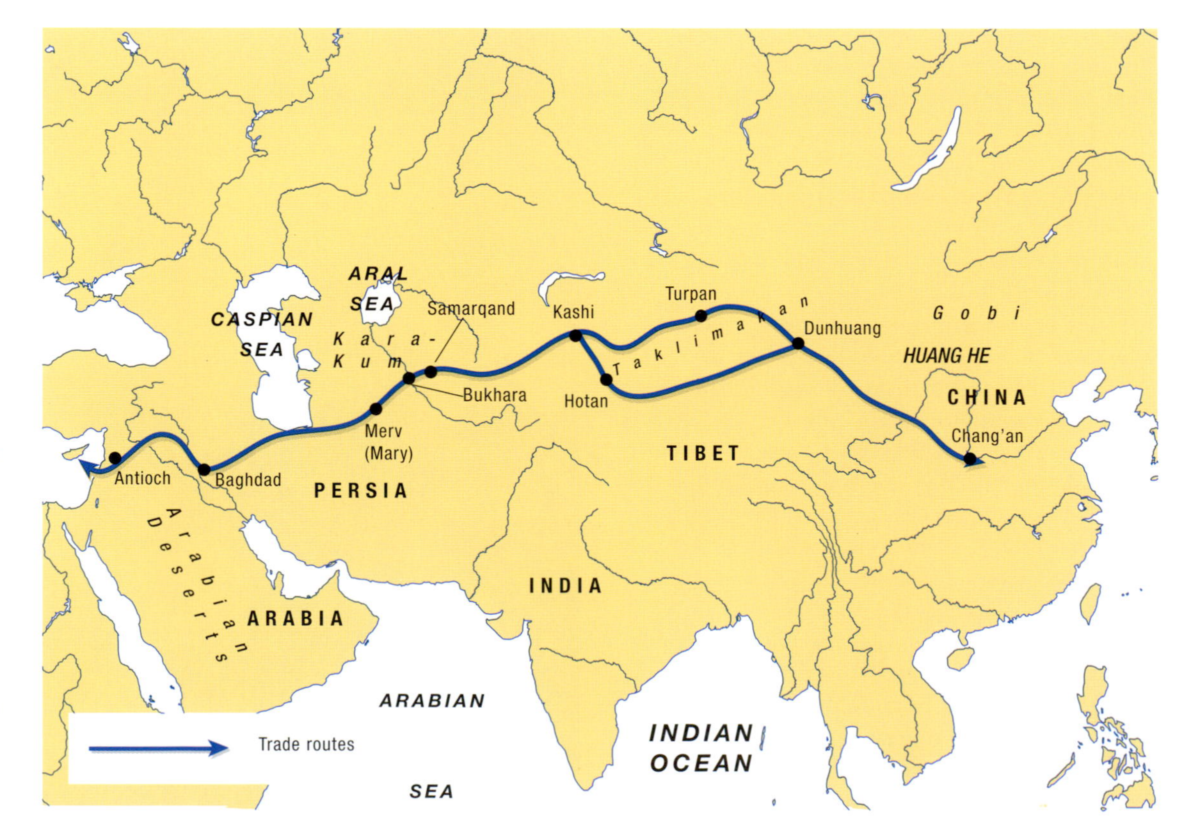


Figure 2 The famous Silk Road was in fact a network of trans-Asian routes, which changed considerably over the centuries. There were, however, two main "roads," both of which started at the Han capital, Chang'an (Xi'an). Both routes skirted the dreaded Taklimakan Desert, which, in addition to the natural hazards of heat, drought, and sandstorms, was haunted by bandits.

**Among the most successful of the desert peoples of Asia were the nomadic Mongols, whose homelands lay on the margins of the vast Gobi Desert. Their first great leader, Genghis Khan, depicted here by a 16th-century Persian painter from Bukhara, founded an empire that stretched from the Danube to the Pacific Ocean. Genghis Khan (top) and his four sons (right) are shown in heavily Persianized guise.**

who brought disease and misfortune, and spirits who became present during medicine dances, when some individuals went into a trance and acquired healing powers. Examples of San art survive in vivid rock paintings, the most recent executed by the !xam (also remembered as storytellers), who became extinct in the late 19th century.

## Peoples of the Asian Deserts

The deserts of Asia, from the Kara-Kum to the Gobi, along with its mighty mountain ranges, form a series of daunting barriers to human traffic. Yet both mountain and desert have yielded surprisingly often to determined individuals and even entire peoples on the move. In 138 B.C. the Chinese general Chang Ch'ien (d. 114 B.C.) set out with a caravan of 100 men on the first transcontinental crossing, and reached deep into Central Asia. His achievement opened the way for the establishment of the so-called Silk Road, a network of trade routes that eventually linked the Chinese Han and Roman empires.

Caravans transported Chinese silks, bronzes, spices, and ceramics to the West and brought back gold and silver. The wares changed hands many times, usually at oasis cities, from fabled Bukhara, Kashi, and Samarqand to the westernmost Chinese outpost at Dunhuang, at the eastern edge of the Taklimakan. The oasis towns prospered handsomely from the trade, extracting large profits from the goods handled within their walls.

The Silk Road encouraged the flow of ideas as well as commerce, enabling Buddhism to reach China and Japan, and intrepid Chinese scholars to study Buddhism in its Indian birthplace. Numerous monasteries and sacred grottoes were established along the routes, some of the most beautiful on the edges of the Taklimakan. The hills surrounding the desert are mostly of sandstone, which could be easily quarried or excavated. There was no shortage of funds for the work, particularly from wealthy merchants who were anxious to invoke divine protection or give thanks for a safe desert crossing.

The "silk road" was actually a network, with two principal routes – northern and southern – and a number of branches, including one over the Pamirs to India. Both major routes skirted the dreaded Taklimakan Desert. The Chinese called the desert *Liu Sha*, or "moving sands," because its dunes constantly shifted. Oases in the Taklimakan were few and far between, and thirst and frequent sandstorms claimed many lives. At the heart of the desert was the vast Tarim Basin, which since the end of the last ice age had gown progressively drier as glacier-fed streams dried up or changed course. By the eighth century many formerly flourishing oasis towns and monasteries had been deserted, often to be lost under the roving sands.

The accessibility of the Silk Road, however, was effected less by the natural hazards of desert and mountain than by human hindrances. In the seventh and eighth centuries Arab invasions of Central Asia largely put an end to the East–West trade. In Central Asia fierce Altaic-speaking nomads ranged far and wide, the most feared of whom were the Mongols. In 1206 they were united by Genghis Khan (*c.*1162–1227), and within two generations they had conquered an empire that stretched from Poland to China. One paradoxical result of Genghis Khan's and his successors' bloodthirsty triumphs was a continent-wide "Mongol Peace" that again made the Silk Road viable, enabling Marco Polo (*see panel*) and other Europeans to observe the Mongols and the wonders of China.

Every year the Mongols who lived in China attempted to retain their toughness by returning to their homeland and distinctive way of life: dwelling in *gers*, circular tents made by stretching layers of greased felt over a wooden frame; practicing pony-riding; and living on mutton and koumiss, the native drink made of fermented mare's milk. In 1368 the Mongols were driven from China and their Asian empire fragmented, though formidable Mongol states survived into the 17th century. This hardy people remained widely spread over what are now Mongolia and Chinese-ruled Inner Mongolia, carrying on a traditional nomadic way of life on their high, cold plateau and in the still more extreme conditions of the Gobi.

## MARCO POLO

Europeans learned of the hardships and hazards of Central Asian travel through the Venetian merchant Marco Polo (*c.*1254–1324), who reached China and served the Mongol emperor Kublai Khan (1215–1294), the grandson of Genghis Khan. Marco was a teenager when he set out for East Asia with his father and uncle. Traveling across land from the port of Acre late in 1271, the Polos passed through Persia and Afghanistan, joined the Silk Road at Kashi (Kashgar), and followed the northern route along the edge of the Taklimakan Desert. After resting at the city of Lop, they set out on a 30-day journey across the "Lop [Nur] Desert," which consisted "entirely of mountains and sand and valleys." A series of watering places made survival possible, but the danger of being led astray by strange noises and "spirit voices" prompted the travelers to huddle close together until they reached Dunhuang and the safety of the Great Wall.

Marco's adventures were only recorded because he took part in a conflict between Venice and Genoa after his return to Italy. While he was a prisoner of war, he told his story to a fellow-captive and professional writer named Rustichello; and Rustichello wrote the whole account down – with some embroidery – as *The Travels of Marco Polo.*

## DESERT PEOPLES OF AUSTRALIA

The first human beings reached Australia about 40,000 years ago, when water levels were lower and island-hopping voyages from Southeast Asia could be made without losing sight of land. These Aboriginal Australians settled all over the continent, probably entering its arid interior last of all, around 26,000 to 30,000 years ago. The notion that all Aborigines were primarily desert-dwellers arose after European settlers had driven them from the more fertile regions of the continent.

The Aborigines of the desert were nomadic hunter-gatherers, comparable in many ways to the San of the Kalahari. They acquired an encyclopedic knowledge of the land and its resources, living in small groups and traveling within a well-defined hereditary territory from one source of food and water to the next. The males hunted kangaroo, wallabies, and other desert mammals, using wooden spears and spear-throwers, boomerangs, stone knives, and possibly the dogs that are the ancestors of the modern dingo (*see* p. 109); the women and children collected vegetables, fruit, eggs, and nutritious grubs. The Aborigines' possessions were few: they wore no clothes, built simple branch shelters when no other protection was available, and, apart from weapons, used a few implements such as digging sticks, wooden dishes, bark or grass baskets, and fiber ropes.

This painting from Nourlangie Rock, an Aboriginal sacred site in the Northern Territory, is typical of artists' concern to portray the natural world around them.

This simple material culture was combined with a rich and complex social life, religion, and art. Aborigines lived by unwritten rules, without recognized leaders, and within a complicated system of kinship and totemic groups identified with one or more animal species. Oneness with the group was felt to be so complete that individuals always married outside it. Rites of initiation, especially the passage from boy to man, involved painful ordeals, such as circumcision and the extraction of teeth. Sociability was promoted by a way of life that left time for leisure, and occasionally groups came together to hold a corroboree, where stories were told and the men, with their bodies painted, danced to rhythmic stick-beats and the music of the didgeridoo.

Aboriginal social customs were part of an all-encompassing spiritual or mythic outlook with its source in the Dreaming, or Dreamtime, when mythical ancestor figures moved across the land, creating all things. The Dreaming was not, in the Western sense, in the past nor a visionary experience during sleep, but was a coexistent dimension that it was possible to enter, among other ways, by maintaining sacred sites and renewing ancient rock paintings and engravings. Aborigine art, ranging from abstract designs to stick and "X-ray" figures (showing stylized internal organs), has impressed the outside world with a sense of the vitality and intensity of a once-despised culture.

## ULURU

The massive monoliths – isolated outcrops of weathered rocks – of the Central Australian deserts play an important role in the religious beliefs of several Aborigine peoples. The best-known of the rock formations is the roughly oval Uluru, also known as Ayers Rock, which rises some 335 meters (1,100 ft.) above the surrounding desert plain and is 3.6 kilometers (2.2 mi.) long by 2 kilometers (1.5 mi.) wide. The rock is formed of arkose – a coarse-grained sandstone – and changes color according to the altitude of the sun. Uluru is especially spectacular at sunset, when it glows a fiery orange-red. The top of the monolith is riven by gullies, which after rare rainstorms produce large waterfalls. Many areas of Uluru are sacred Aboriginal sites, with separate places for men and women, and many Dreaming stories converge at the rock. In 1985 ownership of Uluru was returned to the local Aboriginal peoples, but only on condition that they immediately lease back the land to the Australian government as the Uluru National Park. Every year thousands of tourists make the trip to Uluru from Alice Springs, some 450 kilometers (240 mi.) to the northeast.

## American Desert Peoples

Human beings entered the Americas in relatively recent times – perhaps as few as 15,000 years ago – but by the first millennium A.D. they had spread to populate both continents and created a great diversity of cultures. Of North America's desert regions, the Great Basin was bleak, and the vast Mojave Desert offered only the barest subsistence to scattered bands of Shoshone. There were, however, remarkable developments in the deep Southwest, where cultures such as the Hohokam, the Anasazi, and the Mogollon founded permanent settlements and lived by combining corn- (maize-) growing with gathering and hunting.

The southernmost group, the Mogollons, lived in the mountain and desert country on the modern U.S. border with Mexico. They made high-quality pottery, notably Mimbres ware (A.D. 1000–1130), which was ritually "broken" by being pierced in the base and buried with its owner after death. The Hohokam established farming villages in Arizona, which were supported by extensive irrigation systems. They seem to have been influenced by the civilizations of Mexico to the south, building on earth platforms and constructing ball courts like those in which Mexican peoples played a sacred game, very roughly comparable to basketball, as part of their religious rites.

The Anasazi culture has left behind the most spectacular remains. Living in the arid "Four Corners" region – at the junction of Colorado, Utah, Arizona, and New Mexico – they nevertheless became proficient farmers, growing enough corn (maize), beans, and cotton to establish permanent settlements. Early settlements, or pueblos (from the Spanish for "villages"), were made from adobe blocks, but in about A.D. 1050–1300 the Anasazi, or Pueblos as they are known at this stage of their cultural development, began to build in stone, creating cliff dwellings, often sited in out-of-the-way canyons or on the tops or cliffs of mesas.

Each pueblo was like a multistoried apartment house or palace, sometimes with as many as 1,000 rooms, and was home to a whole community. Each floor was set back from the one below, creating terraces that provided valuable outdoor living space. The ground floor had no outside windows or doors and was used to store grain. One of the most striking features of these sites were semi-subterranean chambers known as kivas where the men met for ceremonial purposes, entering by a ladder through the roof. Among the most spectacular pueblos are Pueblo Bonito in Chaco Canyon, New Mexico, and Cliff Palace at Mesa Verde, Colorado.

Around 1300 the Anasazis abandoned the cliff pueblos, perhaps driven away by prolonged drought, overexploitation of the environment, or the incursion of nomadic Navajo and Apache into their territory. However, their dispersal did not mean the end of the Pueblo culture, which continued to flourish in eastern Arizona and New Mexico. When the Spanish arrived in the region during the 16th century, they were impressed enough to describe not only the settlements but the people themselves as pueblos. The Pueblos were overcome by the Spanish – but not easily. Even in defeat they clung to their own way of life. Although many traditional settlements were abandoned in the Hispanic period, some are still occupied today, and their inhabitants continue to practice ancient skills, such as pottery, and to observe religious rites.

**Most of the cliff dwellings at one of the most spectacular Anasazi pueblos, Mesa Verde in Colorado, were built in the mid-13th century and range from one-room houses to villages of more than 200 rooms, as at Cliff Palace. Many rooms were plastered and decorated with painted designs.**

Pueblos, such as this one near Taos, New Mexico, continue to function as working communities in some parts of the American Southwest. Communal bread ovens like the one shown in the foreground are an important feature of such close-knit settlements.

The Apaches were formidable warriors who arrived in the American Southwest about A.D. 1300. They remained nomadic hunters, living in brush shelters known as wickiups. By contrast, the closely related Navajo did settle, living in small groups and making their homes in earth-covered wooden structures called hogans. They learned to farm and weave from the Pueblos, and Navajo blankets and rugs became celebrated for their stylized striped designs. Unconquered by the Spanish and Mexicans, the Apaches and Navajos were eventually defeated by U.S. forces in the later 19th century.

In South America the Atacama Desert in Chile was inhabited only by small groups who managed to irrigate and farm a few oases; nevertheless, influenced by the Inca civilization to the north, the Atacamenos became skilled craftsmen. Farther south, life in the Patagonian Desert was even harder. The best-known of its peoples, the Tehuelches, were nomadic hunters called by the Spanish *Patagones*, "big feet." They gained some advantages from the arrival of Europeans, mounting the horses introduced by the newcomers and for a time maintaining an independent lifestyle.

# THE MODERN DESERT

*Desert life has been transformed in recent times, and the old ways have largely disappeared or have been greatly modified. This was brought about in many places by the arrival of Europeans, and everywhere by the impact of new economic imperatives and technologies.*

---

Many deserts are no longer impenetrable or isolated wildernesses whose ways and secrets are known only to a few hardy inhabitants. Highways and railroads cut through sand dunes; modern cities have grown around rare oases; airplanes connect remote settlements; and industry scars the land wherever resources can be found and mined. Most of these changes have occurred only very recently, as modern technologies have opened up deserts to settlers, speculators, industrialists, and tourists alike.

**The desert city of Riyadh, capital of Saudi Arabia, grew rapidly from the 1950s on the back of oil wealth. New palaces, government buildings, mosques, and suburbs have transformed the old walled oasis town beyond recognition.**

The technological development of the desert was often prepared for by its colonial appropriation by European powers in the 19th and early 20th centuries. Interest in the desert that began as seemingly disinterested scientific curiosity – for example, the search for the source of the Niger or Nile – or as an attempt to spread the Christian message among its peoples often heralded a more far-reaching colonial project. The effect of both colonial and technological encroachments on the desert has had a profound impact on its inhabitants, with many more or less forced to abandon their traditional lifestyles. A few, such as the pastoralists of Mongolia, have managed to preserve their age-old culture while taking what is best or useful from the new.

## European Explorations

From the late-18th century adventurous Europeans increasingly came into contact with the world's desert regions. Their understanding of desert regions sometimes owed more to imagination than reality, however. The reports and travel writings of explorers more often than not fueled rather than dispelled romantic myths about the desert life that lay beyond civilization. In the 19th century painters and writers used the Sahara and Arabia as backdrops for Orientalist fantasies, depicting a world of sensuality and cruelty that had much more to do with the European psyche than with Islamic reality. Only a little later the myth of the American "Wild West" took root, in which "civilized" settlers and "barbarian" natives struggled over frontier lands which included the deserts and semiarid lands of the Southwest.

European explorations of the desert, and the myth-making they helped engender, often went hand in hand with the European colonial enterprise. The "discovery" of new lands, and the gathering of knowledge about them, was in many cases a precursor to invasion. In the European mind, deserts, above all, were considered "no-man's land" and therefore open to annexation.

### To Tombouctou and back

The secrets of the Sahara were jealously guarded by the Tuareg and its other nomad peoples, who were especially hostile to non-Muslims. European interest stirred in 1788, when the African Association in London began to send out expeditions to try to reach the fabled Tombouctou (Timbuktu) and locate the source of the Niger River. Thanks to heat, malaria, and the hostility of the Tuareg, the casualty rate on such expeditions was high.

In 1798, for example, the society commissioned the German explorer Friedrich Konrad Hornemann (1772–1801) to make a further attempt. Having first taken the precaution of disguising himself as a Muslim, Hornemann set out from Cairo, traveling with a caravan via the Siwa Oasis to Mursuq, in present-day Libya. From Tripoli the intrepid traveler sent home his records and maps to his employers in London. No accounts of his subsequent travels from Mursuq via Bornu and Lake Chad to Niger are known. Hornemann is said to have died in 1801 at the Niger River. The account of his journey, published later, contains the first modern European descriptions of the western Sahara, which was previously unknown.

The Scottish explorer Hugh Clapperton (1788–1827) traveled much farther south, reaching Kano, the Hausa capital, and Sokoto, capital of the powerful Fulani Empire. It was another Scottish explorer, Alexander Laing (1793–1826), however, who became the first modern European to reach Tombouctou. He spent five weeks in the town, but was murdered as he left with a caravan. A young Frenchman, René Caillié (1799–1838), became the first European to reach Tombouctou and live, though he found there nothing of the city's former glory. Caillié had walked all the way from Sierra

## NAPOLÉON IN THE DESERT

One of the most extraordinary episodes in the career of Napoléon Bonaparte (1769–1821) was the French conquest of Egypt in 1798. In launching the expedition, Napoléon hoped not only to cut off one of the trade routes between France's principal enemy, England, and its lucrative territories in India, but also to win some easy glory that would deflect attention away from the faltering regime at home. Once the expedition was launched, the French public was highly enthusiastic; for many French people, Egypt was a land of mystery and exoticism – an image popularized in late-18th-century travel books, such as Volney's *Les Ruines* (The Ruins) – and Napoléon a new Alexander.

The expedition was organized within a month, and with little understanding of the rigors of the Egyptian terrain. On May 19, 1798, a fleet of 200 ships carrying some 35,000 men left Toulon and set sail for Egypt. On board were not only soldiers and other military personnel, but representatives of the arts and sciences, including 21 mathematicians, 3 astronomers, 17 civil engineers, 13 naturalists, as many geographers, 8 draftsmen, 10 writers, a sculptor, an engraver, a poet, a flower painter, and a pianist. Napoléon's intention was to grace what was essentially a political expedition with a scientific gloss. The expedition set out not only to conquer Egypt but to catalog it, and the two campaigns – military and scientific – went hand in hand.

The morale of Napoléon's army rapidly declined after the fleet landed at the Egyptian port of Alexandria on July 1. Confronted with Egypt's deserts, rather than with the Orientalist fantasies pedaled in France, many lost faith in their leader. One observer recorded: "Our soldiers were dying in the sand from lack of water and food; an excessive heat had forced them to jettison their spoils, and some, weary of suffering, had blown out their brains."

Napoléon's expedition seemed doomed from the start. Napoleonic Egypt found itself under attack not only from the outside – as the British and Ottomans laid siege to the country from land and sea – but from the inside, as Egyptian uprisings broke out in Cairo. By 1801 Napoléon had agreed to an armistice and retreated from the country. The one great achievement of Napoléon's desert adventure was the monumental 24-volume *Description de l'Egypte*, in which the country's land, people, flora, and fauna, together with its ancient monuments, were described, annotated, and illustrated for posterity.

In Europe the Muslim desert pilgrimage to Mecca was the object of much romantic speculation. In 1861 the French Orientalist painter Léon Belly, who had traveled widely in the Middle East, exhibited this large desert painting entitled *Pilgrims Going to Mecca*. Unlike his Romantic precursors such as Delacroix, Belly was concerned to give an ethnographically exact depiction of a pilgrimage.

Leone on Africa's west coast and returned to Europe across the Sahara in order to claim the 10,000-franc reward offered by the French Geographical Society. Scientific investigation came to the fore during a British expedition of 1850 to 1855, whose German leader, Heinrich Barth (1821–1865), reached Tombouctou and made records of the vast landscapes through which he traveled.

At the other end of the continent the Kalahari Desert was known to Dutch pioneers venturing north from the Cape as early as 1760. The first Europeans to cross the desert were the famous Scottish explorer and missionary David Livingstone (1813–1873) and his English friend the hunter William Cotton Oswell, traveling northward from Livingstone's mission station at Kolobeng to Lake Ngami, in present-day northwest Botswana.

### The lure of Arabia

European exploration of the arid Arabian Peninsula was hampered by the hostility of the Muslim population, bedouin and townspeople alike, to intrusions by non-Muslims. For some Europeans this danger only heightened the challenge. In 1503 an Italian adventurer, Ludovico di Varthema, became the first known European to take part in the annual Muslim pilgrimage to Mecca, having enrolled under the name of Yunas in a Mamluk garrison in Damascus.

According to di Varthema, who later wrote an account of his travels, the caravan that journeyed from Damascus to Mecca encompassed 40,000 pilgrims and 35,000 camels, along with an escort of 60 Mamluks. The caravan's first stop inside the Hejaz (the western region of Arabia and the seat of the Muslim holy city) was at the Jewish oasis settlement of Khaybar, whose inhabitants di Varthema described as "five or six spans tall ... more black than any other color, [living] entirely on the flesh of sheep and nothing else." Others, with more scholarly credentials, followed in di Varthema's footsteps. These included the Swiss discoverer of the rock temples of Abu Simbel, Johann Ludwig Burckhardt (1784–1817), who crossed the Nubian Desert on his way to Mecca; and the Briton Sir Richard Burton (1821–1890), the great Victorian Orientalist who later searched for the source of the Nile. To visit Mecca, Burton disguised himself as a Pathan, or Afghan Muslim. He later published a perceptive account of his experiences in *Pilgrimage to El-Medinah and Mecca* (1855–1856).

The hostile interior of Arabia was only slowly penetrated in the 19th and 20th centuries, mainly by a series of British scholar-travelers like Burton,

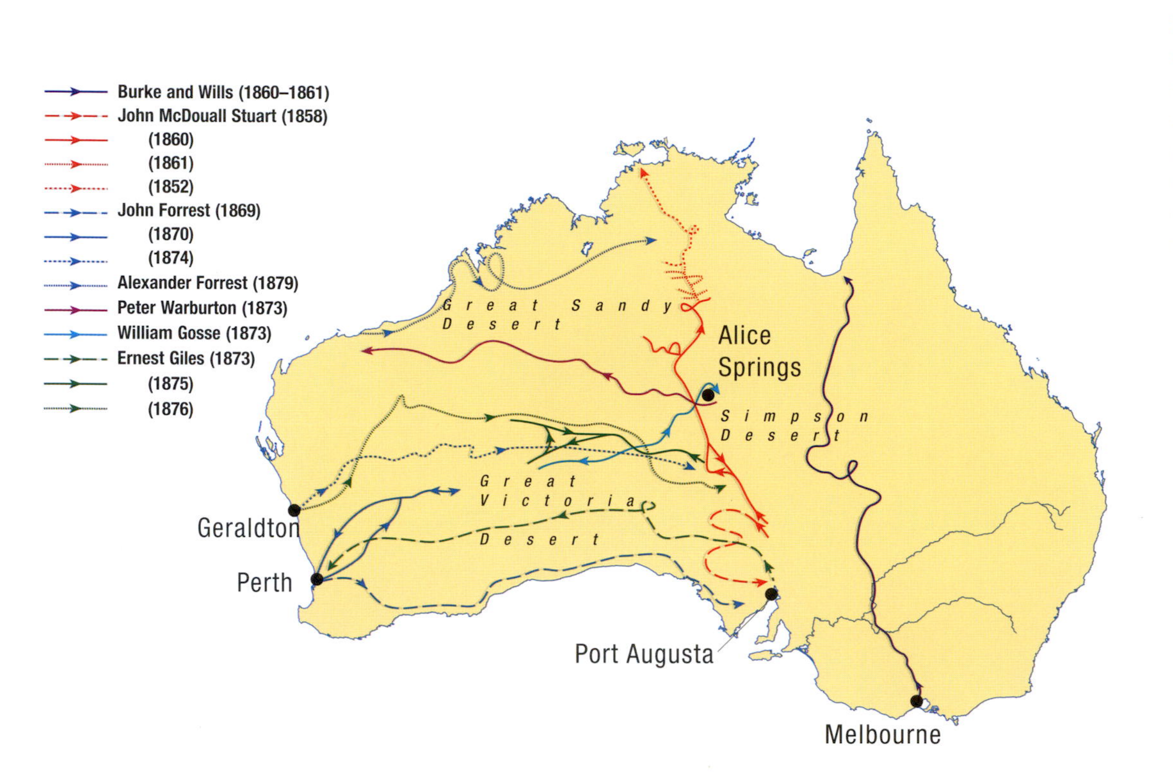


**Figure 1** **In the late 19th century expeditions set out to explore Australia's arid interior. Many explorers died or came close to death, but their persistence enabled hitherto isolated areas to be colonized.**

who often formed a deep attachment to the desert and its bedouin inhabitants. In 1862 William Palgrave (1826–1888) and a Greek companion crossed the An Nafūd Desert in what is now northern Saudi Arabia, braving its burning simoom wind and mountainous red dunes, and became the first Europeans to travel across the Arabian Peninsula. In the late 1870s Charles Montagu Doughty (1843–1926) and Wilfred Scaven Blunt (1840–1922) traveled extensively in the interior, living among bedouin peoples. It was not until 1930 to 1931 that the Empty Quarter (*see* pp. 40–41) was crossed by Bertram Thomas, an adviser to the sultan of Muscat, just before the feat was also accomplished by Harry St. John Philby (1885–1960), who also became the first European to visit the southern provinces of the Nejd. The lure of the Empty Quarter remained so strong that in the late 1940s it was extensively traveled by another British traveler, Wilfred Thesiger (born 1910), whose *Arabian Sands* includes a classic account of bedouin life.

Much of the deserts, mountains, and steppes of Central Asia was mapped by Russian pioneers in the course of extending their empire during the 18th and 19th centuries. The most distinguished European explorer of the region, however, was the Swedish geographer Sven Hedin (1865–1952), whose 40-year travels (1893–1933) ranged from Persia to China and included the fearsome Taklimakan Desert. "The whole of Asia was open before me," Hedin once wrote, "I felt that I had been called to make discoveries without limits – they just waited for me in the middle of the deserts and mountain peaks ... my first guiding principle was to explore only such regions, where nobody else had been earlier." On a different timescale the American Roy Chapman Andrews (1884–1960) was the first Western explorer to discover that the interior of the Gobi Desert was a vast repository of dinosaur and other ancient remains.

## BURKE AND WILLS

After strenuous lobbying, Robert Burke (1821–1861), an Irish-born police inspector, was chosen to lead an expedition across Australia from south to north; William Wills became his second-in-command. The expedition left Melbourne in August 1860 with 25 camels and ample stores. Burke set up a depot at Menindee, on the Darling River, and then divided the party, going ahead with seven men to Cooper's Creek, some 600 kilometers (370 mi.) to the northwest. He waited until mid-December for the rest to catch up and then, losing patience, divided the party again. Burke, Wills, John King, and Charles Gray went forward; those left behind were instructed to wait up to three months for Burke's return.

Burke was fortunate in the weather and reached the Gulf of Carpentaria as planned, but the journey had taken two months and provisions were almost exhausted. The return journey became a terrible ordeal as the men's strength failed and the camels perished. Gray died, but the others reached Cooper's Creek on April 21, 1861, to find that their companions had given up and gone only seven hours earlier. Burke, Wills, and King wandered for two months, again narrowly missing a chance of being rescued, before first Wills and then Burke died. King survived on seeds and fish supplied by Aborigines; when a rescue expedition found him, he was a living skeleton.

Native Aborigines thrived in the desert area surrounding the giant monolith known as Uluru for thousands of years before Europeans reached the continent. The first European to reach and climb the rock was William Christie Gosse (1842–1881) in 1873. Gosse named the monolith after Sir Henry Ayers, a South Australian government leader.

## Desert explorers of the "New World"

The first recorded Europeans to visit the American Southwest were Álvar Núñez Cabeza de Vaca (*c.*1490–*c.*1560) and his three companions – Spaniards who had wandered for years after being shipwrecked and who eventually stumbled into Zuni territory in 1536. The tales they heard – or imagined they heard – of seven cities of gold lying farther north – whetted the appetites of the Spanish authorities in Mexico, who in 1540 sent out a northern military expedition to seek the fabled treasure. The expedition found no gold but marked the beginning of the oppressive Spanish domination of the Pueblo peoples.

Americans' contact with the region began only in 1828, when a trapper, Jebediah Smith, crossed the Great Basin. Joseph Walker and John Frémont conducted extensive surveys in 1833 and 1843 to 1845 respectively, and parts of the region became notorious among pioneers heading for California who suffered the predations of the deserts and mountains and fell victim to heat, cold, or starvation (*see* p. 20). The deserts of South America were also discovered by 16th-century Spaniards who sought, but failed to find, riches comparable to those they had found in the plundered Inca Empire of the Andes.

Once European settlements had been established on the southeast and western edges of Australia by the early 19th century (a slow process that had taken hundreds of years), hopes of linking the two spurred efforts to cross the interior. Many expeditions were forced back by the deserts in the heart of the continent, and many explorers perished. From 1846 to 1847, for example, the Prussian Ludwig Leichhardt (1813–1848) was forced by heat

**Desert lifestyles today often blend the old and the new. Here bedouin women of the Negev Desert, Israel, play with a child in a tent decorated with mass-produced carpets.**

and drought to turn back from an attempt to traverse Australia from east to west. Shortly afterward he again set out on an overland journey to Perth. From this expedition he did not return. In 1859 the offer of a £2,000 prize for the first south–north crossing prompted two competing expeditions, both of which set out in 1860. John McDouall Stuart (1815–1866), already a veteran explorer, was forced to turn back but did at least achieve an earlier ambition, planting the British flag in the dead center of Australia. By contrast, Robert Burke and William Wills (*see panel*, p. 148) achieved their goal but died on the return journey. In 1861 and 1862 McDouall Stuart crossed the continent from Adelaide to reach the northern coast close to the site of the future city of Darwin. Stuart's route became the basis of the Overland Telegraph Line, completed in 1872, and of the original Central Australia Railway (Ghan) from Adelaide to Alice Springs. The telegraph line's relay stations, established at regular intervals across the continent, facilitated further exploration, and in the 1870s Ernest Giles, Peter Warburton, and John and Alexander Forrest crisscrossed the interior of Western Australia, putting its vast uninhabited desert interior on the map.

## Desert Peoples Today

Exploration in many cases preceded large-scale European intrusion into desert regions, and the lifestyles of desert peoples were often irrevocably changed. Pressure on desert peoples has increased relentlessly in more recent times. Outsiders have moved in, armed and aggressive, carrying new diseases, or, at best, peacefully competing for already scarce resources. The deserts have attracted industry, mainly to extract their mineral resources (*see* Chapter 7), and modern states, with their impulse to control, have tried to tempt or compel nomads to adopt a settled existence. Moreover, since the nomad's life is a hard one, the temptations of conveniences and luxuries have made modernity enticing to many, especially when traditional ways have become even more precarious through the impact of war, climatic change, and tourism.

### Deserting the desert

The nomads of the Sahara retained their independence even after the French conquest of Algeria in the 1840s. In the 1880s and 1890s the Tuareg were able to defy the colonial power, and it was only in 1902 that the French, having enlisted the Tuareg's enemies on their side, were able to defeat them decisively. Though no longer lawless, the Tuareg remained free to roam the desert.

The states that replaced the French in the 1960s proved far more determined to make their nomadic subjects – whose lifestyle ignored political boundaries and evaded political control – settle down. Circumstances favored them, as competition for precious oasis soils intensified, the camel became uneconomic by comparison with motorized vehicles, and, above all, prolonged droughts from the 1970s made living in the desert increasingly difficult and untenable. Some nomads adapted, adapting the plastic tent and the truck to their traditional lifestyles, but large numbers also sought work in oasis towns or mushrooming cities.

Similar developments drastically reduced the numbers of nomadic bedouin in the Arabian Peninsula. A policy of urbanization was adopted as early as the 1900s by the Arab leader Ibn Sa'ūd (*c.*1880–1953), who found the bedouin too independent and therefore unreliable, and appeared to be politically justified when his victories led to the creation of Saudi Arabia. The discovery of oil created employment and a higher standard of living for many people in Saudi Arabia and the Gulf States. Only very small numbers continued to herd flocks or trade in commodities such as rock salt.

### Desert poverty and desert wealth

Some desert peoples have lost the old without benefiting much from the new way of life. The San of southern Africa have suffered from the encroachments of white ranchers and the Bantu-speaking Herero people of Namibia, mining operations in the Kalahari, drought, government-imposed restrictions on hunting, and the long armed struggle in Namibia. Now only a few San follow something resembling the traditional way of life in the Central Kalahari Game Reserve; many more form an underclass working for the farmers and ranchers who have displaced them.

The treatment of Aborigines in Australia was even worse, owing to the racist attitudes of many settlers. Those who were not killed or who did not succumb to unfamiliar diseases were driven from their territories by ranchers and miners, on the quasi-legal grounds that Australian land was *terra nullius* ("no-man's land"), without owners before the arrival of Europeans. Aborigines ended up in ramshackle settlements or in city slums, often unemployed, in poor health, succumbing to alcoholism, and viewed by other Australians as a problem rather than a responsibility.

Many native peoples in North America were suffering a similar fate even before the Indian Wars ended with the surrender of the Apache leader Geronimo (1829–1909) in 1886. For a time the decline in Native American populations gave rise to the belief that they had become the "Vanishing Americans," doomed to assimilation or extinction. Many tribes were uprooted and sent to reservations far from their homes, where they could not follow their traditional pursuits. The Indians of the Southwest, however, were not compelled to forsake the desert, if only because desert territory was regarded as being of little value.

In the case of the Navajo this was a conscious decision, made in 1868 when they chose to stay rather than move to lusher land in Oklahoma. At all times Native Americans in the Southwest remained poorer than most Americans. Pueblos who wanted to stay in their villages were able to observe traditional customs and ceremonies; the western Apaches adapted to livestock raising and farming;

## CAMEL RACING

Racing camels has long been a popular pastime, and not just in their traditional homelands. After camels were introduced into the United States in the mid-19th century, races were held every year at Virginia City; and nowadays Maralal in Kenya has its own Camel Derby and Alice Springs in Australia its International Camel Cup. The most competitive and professional events, however, are held in Saudi Arabia and the Gulf States, where wealthy city-dwellers own mounts in the way that people elsewhere own racehorses, employing professional jockeys 11 to 15 years old. Camels are specially bred, some by artificial insemination, and the speediest, capable of 20 kilometers (12.5 mi.) an hour or even more, command very high prices. In Saudi Arabia thousands compete in the King's Camel Race, and there is a regular program of events at Ra's al-Khaymar in Dubai.

and the Navajo began to acquire a reputation as makers of pottery, baskets, rugs, and beautiful turquoise and silver jewelry.

Life in the desert is associated with the rigors of nomadism, or with the easier, but still strenuous, existence of the oasis-dweller. In the 20th century, however, the deserts yielded immense wealth, creating a super-rich class. Exploitation of Middle Eastern oil began in the early 1930s, but its real impact on Saudi Arabia and the Gulf States dates from the 1950s. Though prices have fluctuated, peaking during the oil crisis of 1973 and falling off after 1985, revenues over the decades made the rulers of the region fabulously rich while also bringing benefits to ordinary citizens.

Oil wealth supported the rule of the shah of Iran, Mohammad Reza Pahlavi (1919–1980), but did not prevent him from being overthrown in 1979 by the Islamic Revolution. Despite the fundamentalists' rejection of capitalist values, oil income remains vital to many Islamic economies. Oil revenues funded the ambitions of Saddam Hussein

## STRANDED IN THE DESERT

Modern technology has made it possible for the tourist or would-be adventurer to travel deep into the desert without guides or previous experience. Sturdy four-wheel-drive vehicles and mobile phones seem to offer a high degree of security. All the same the desert remains a dangerous place, where a breakdown is a serious matter and sudden storms may leave tracks invisible under a blanket of sand or dust. Apart from obvious precautions such as carrying compasses, a first-aid kit, equipment for repairs, and more-than-adequate supplies of food and water, the traveler can follow the examples of the professionals whose trucks make regular journeys across the Sahara. They carry spades to clear the sand from around wheels and mats that are placed in front of the wheels so that they have something to grip on when a vehicle is being released from a sand trap. Clothing that will protect the skin from the sun is essential on such occasions – and all the time for bikers. A long drive becomes less punishing when the first stage is made at night, and when traveling significant distances along unpaved tracks, the soundest precaution of all is to travel as part of a convoy, so that mutual aid is available throughout the trip.

The Pan-American Highway cuts its way through the windy Sechura Desert on the Peruvian coast. Today many hitherto largely impassable deserts are crisscrossed by highways and railroads.

(1937–), dictator of Iraq, and helped him to survive a disastrous defeat in the Gulf War. The state invaded by Saddam's troops, Kuwait, became notable for importing so many servants and workers from abroad that its own citizens formed a privileged minority of the population.

In Saudi Arabia, Riyadh, the capital, grew from an oasis city to a metropolis of two million people; this and other urban centers thrived thanks to expensive desalination schemes. Large-scale building projects were undertaken in the peninsula, ranging from mosques to office blocks and golf clubs. Less significant but better publicized were the displays of great wealth – private jets, fleets of automobiles, wildly extravagant lifestyles, acts of generosity, and, sometimes, scandalous doings – that identified those who indulged in them as kings of the – oil-rich – desert.

## Tourists and transportation

Technology has opened up most deserts, and affluence has made it possible for an increasingly large number of people to visit them. Airlines take tourists to the remotest places, where they can be guided into, or even across, deserts, traveling by camel or mechanized transport. Those with the time, money, and inclination can go on safari in the Kalahari, ride a camel out of Alice Springs, stay in a tent, bedouin-style, in the Arabian Desert, or follow the Silk Road all the way by train.

Though most indigenous peoples of the desert are under threat, many new inhabitants have arrived in recent times. Farmers in Israel's Negev Desert and elsewhere grow crops in artificial conditions, making highly efficient use of available water (*see* p. 173). Miners extract diamonds from the Kalahari, oil from the Arabian Peninsula, and copper from the Atacama. In Saudi Arabia and the United States entire cities are sustained by installing desalination plants or piping water from underground or distant sources. Grandiose Las Vegas stands in the heart of the Nevada Desert, while Phoenix in Arizona attracts, among others, large numbers of retired people. Climate aside, both cities might be located anywhere in modern America.

More people than ever before travel in the desert, as tourists, traders, or workers, and although camels are still used, they have been replaced for most purposes by mechanized transport. The first automobiles to cross the Sahara were five Citroëns equipped with giant headlights and front tires and specially designed rollers instead of back wheels. They set out from Touggourt in Algeria in December 1922 and reached Tombouctou 20 days later, beating the time of even the fastest camel by over two months. The scientist Roy Chapman Andrews

In the arid U.S. Southwest lush green spaces are often sustained using elaborate methods of irrigation. This golf course in Page, Arizona, uses recycled water from the city.

used Dodges to equal effect during his Gobi expeditions of the 1920s.

Since that time parts of the Sahara have become busy thoroughfares, with trackside markers indicating the main routes, and trucks and automobiles steadily heading north or south. More affluent desert travelers use four-wheel-drive vehicles with big, sand-gripping tires to make journeys, explore, or race. In some countries regular roads cross the desert, and on Saudi Arabian roadsides solar-powered telephones are placed at regular intervals so that the traveler can summon assistance in an emergency. Driving through the desert for sport has become part of the thrill of rallying, featuring on routes such as Paris to Dakar and Istanbul to Beijing.

Mountain bikes, motorbikes, and even microlites have invaded the desert, but the oldest form of mechanized transport, the train, crosses it so swiftly and efficiently that new lines have continued to be opened. In the 1950s the Trans-Mongolian line from Beijing to Ulaanbaatar was completed, crossing the great Gobi Desert and linking the Chinese and Mongolian capitals. The trans-Australian Perth to Sydney line, opened in 1970, traverses the desert Nullarbor Plain by the longest stretch of straight track in the world (462 kilometers or 287 miles). And in 1990 the opening of the Ūrūmqi–Alma-Ata line completed the railroad network that forms the modern equivalent of the famous Silk Road.

## Reclaiming the desert

In the later 20th century some native peoples began to reassert their rights and attempted to regain lost territories. This was especially true of the Native Americans and Australia's Aborigine peoples, for both of whom the land has a special, sacred meaning. Both, too, benefited from changing attitudes among the majority population, especially in the 1960s, and from the example of the black and other civil-rights movements in the United States.

Native Americans in the United States were generally allocated the worst lands, and were not secure on them if they became economically valuable for their minerals or tourist potential. One example in the Southwest occurred in 1906, when the Taos Pueblo Indians were deprived of the area around Blue Lake, which was taken over by the Forestry Commission. Barred from performing their traditional ceremonies on the site, the Indians undertook a series of court actions that finally achieved success over 60 years later, in 1971.

**Traditional desert lifestyles are particularly resilient in Mongolia, where pastoralism remains an important part of the economy. The large tents, or *gers*, used by the Mongolian herdsmen and their families are often elaborate, semipermanent affairs.**

Meanwhile Native Americans became citizens of the United States in 1924 and, instead of vanishing, began to multiply. Despite some legal and social gains in the 1930s, assimilation remained the aim of successive U.S. administrations, and Native Americans were encouraged to migrate to the cities. Their culture proved surprisingly resilient even in urban surroundings, although it was not until the 1960s that militant actions such as the occupation of Alcatraz Island signaled a real reform. In the 1970s Native Americans' right to run their own affairs was recognized, and confidence was boosted by victories such as the Blue Lake decision and the 1975 restoration of Havasupai land in Arizona, which had been confiscated for the Grand Canyon National Park.

Present-day Native Americans in general, and those of the Southwest in particular, are largely able to live as they choose. They are nevertheless faced with many problems, especially where powerful city or corporate interests encroach on their water rights or establish themselves on traditional lands in order to exploit reserves of coal, oil, and other minerals.

In Australia change came still more slowly. Even after World War II (1939–1945) the aim of government policy was the assimilation of the Aborigines, who remained under official control without receiving any concomitant benefits. The Aborigines were finally accepted as citizens in 1967, and by the early 1970s an official policy of self-determination was in place. From 1976 Aborigines were able to claim land in the Northern Territory, but only if it was not owned or leased by anyone else – which in effect meant desert areas. In 1992 the High Court overthrew the doctrine of *terra nullius* and recognized that the Aborigines had owned their lands before Europeans arrived in Australia. The decision and the legislation that flowed from it were bitterly opposed, especially by the mining lobby, but in reality the Aborigines gained only limited rights to assert their title to land or to veto mining and other developments on it. The legal processes in establishing title are protracted and expensive, and currently the only substantial Aborigine holdings – far less than their claims – are in the Northern Territory and South Australia.

Meanwhile, although serious social problems remain, Aborigine culture has undergone something of a renaissance and Aborigine paintings have become so valued by the outside world that forgeries have begun to appear. By the 1990s immigration – especially by non-Europeans – had turned Australia into a multicultural society, generally sympathetic to a 10-year campaign for white–Aborigine reconciliation that culminated in mass marches in December 2000.

**An abandoned Navajo dwelling in the Arizona desert is eloquent testimony to the disappearance of traditional ways of life as America's desert peoples move from the land to cities such as Los Angeles and Kansas City.**

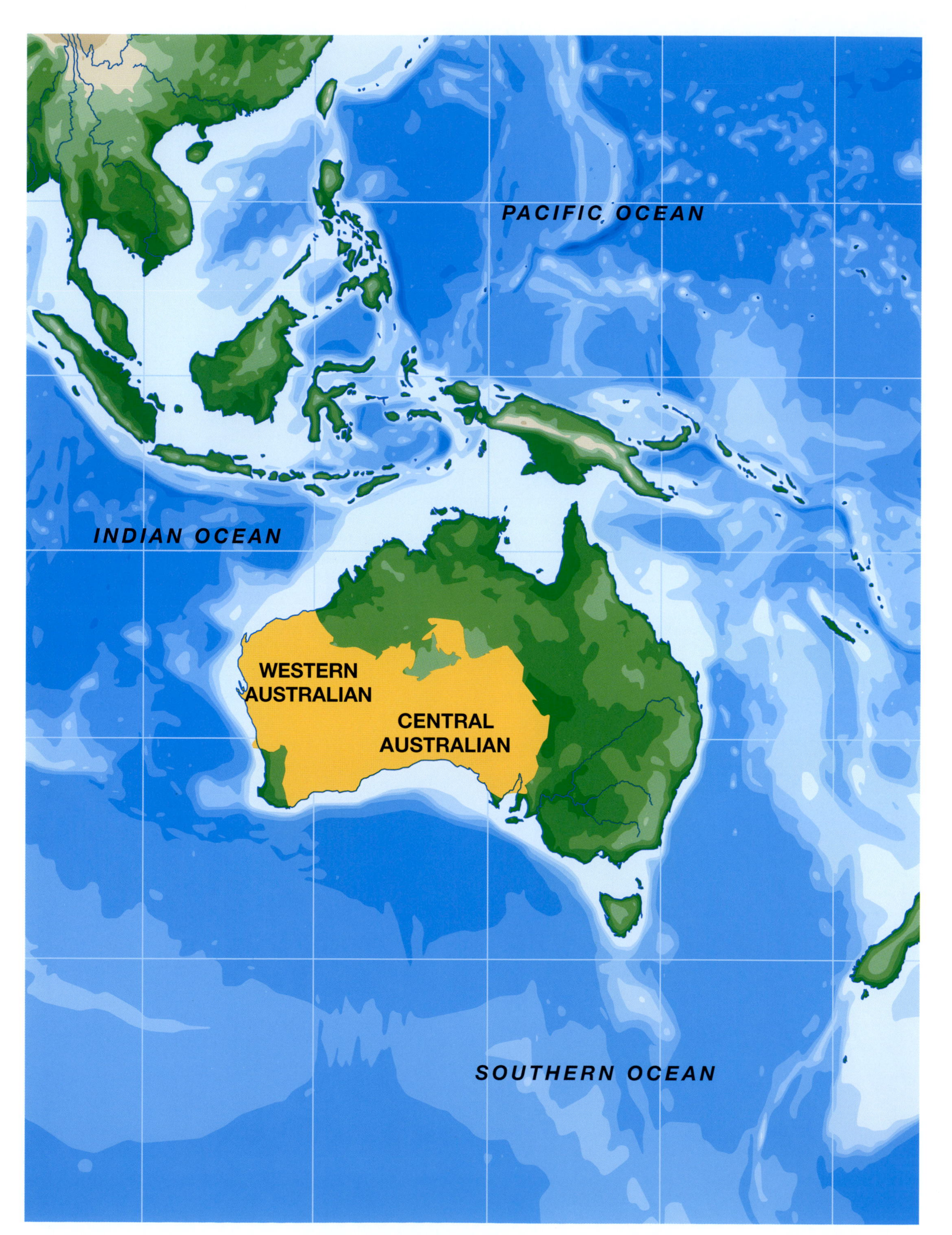
PACIFIC OCEAN
INDIAN OCEAN
WESTERN AUSTRALIAN
CENTRAL AUSTRALIAN
SOUTHERN OCEAN

# AUSTRALIA AND THE POLES

*Inhospitable as Australia's arid interior may appear, it in fact provides a habitat for some of the earth's most extraordinary wildlife as well as a home for many of the continent's first peoples, the Aborigines. By contrast, the "white deserts" of the Arctic and Antarctic are without doubt the planet's ultimate wastelands.*

**Aboriginal women prepare to dance at a corroboree, or informal gathering, in the Australian outback.**

# Western Australian Desert

**1. Purnululu National Park**
One of Australia's greatest wonders is Purnululu, or Bungle Bungle, National Park. It is an area of chasms, gorges, and huge, striped rock domes shaped like beehives. Its 320,000 hectares (1,235 sq. mi.) can be viewed on foot or by helicopter. Excursions to the park depart from the towns of Halls Creek and Turkey Creek.

**2. Halls Creek**
Halls Creek was Western Australia's first gold-rush town in 1885. Four years later the gold was depleted, and many prospectors left for richer pickings in the Eastern Goldfields to the south.

**3. Longest river**
At 290 kilometers (180 mi.), the Fitzroy River, on the Kimberley Plateau, is the longest river in the Western Australian Desert. In dry periods the river consists only of a string of pools, but after a good rain it can flood the land for miles around.

**4. Kimberley Plateau**
This great plateau is often described as Australia's last frontier. It is made up mostly of sandstone, with some volcanic rock, and covers an area larger than that of New Mexico. The only human activity here consists of cattle stations and small, remote communities of Aborigines. The region is prone to raging floods and bushfires. A plan is currently under discussion that would turn all of the Kimberley Plateau into a national park.

**5. Great Sandy Desert**
This vast, empty wasteland of salt marshes and sand dunes dominates northern Western Australia, stretching from the Eighty Mile Beach to the Northern Territory and from the Kimberley Plateau to the tropic of Capricorn.

**6. Marble Bar**
This town (just off map) has recorded Australia's highest ongoing temperatures – 36°C (100°F) or more, lasting for a period of 162 consecutive days. Temperatures as high as 50°C (122°F) have also been recorded here. The town's name derives from a large, marblelike bar of jasper that lies across the nearby Coongan River.

**7. Warburton**
This remote town was named after the first European to cross the Great Sandy Desert – the British explorer Major Peter Egerton Warburton, who made the journey in 1873.

**8. Gibson Desert**
This remote, arid region connecting the Great Sandy and Great Victoria deserts was named after Alfred Gibson, a member of a party of British explorers who

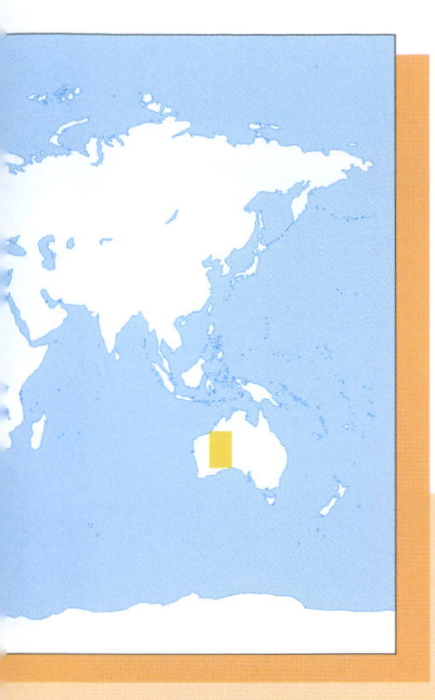

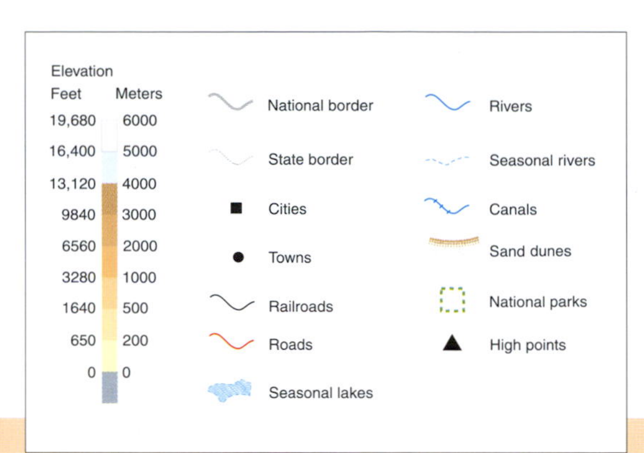


became lost while looking for water in the desert in 1874.

**9. Canning Stock Route**
This 1,600-kilometer-long (1,000 mi.) desert road stretches from Wiluna via Lake Disappointment to Purnululu.

**10. Eastern Goldfields**
Gold was discovered in the Eastern Goldfields at the end of the 19th century, bringing about the settlement of the twin towns of Kalgoorlie and Boulder on the western fringe of the bleak Great Victoria Desert (*see* p. 160). The swell in the area's population caused a severe water shortage, which was solved by a 556-kilometer (345-mi.) pipeline, built in 1903. Kalgoorlie, the more prosperous of the two towns, remains at the core of Australia's gold-mining industry today.

**11. Coolgardie**
This mining town is preserved almost as a museum to the gold-rush days. During its prime in the 1890s the town had 23 hotels, 6 newspapers, and a population 10 times higher than its currently modest 1,500.

**12. Nickel deposits**
Vast nickel deposits were discovered at Kambalda in 1964.

**A lone gum tree (genus *Eucalyptus*) stands in the desert near Warburton, Western Australia. Vast tracts of the Australian outback are dominated by acacias and gum trees.**

# Fact File

- The Western Australian Desert comprises three deserts: the Great Sandy Desert in the north, the Gibson Desert in the center, and the western portion of the Great Victoria Desert in the south.

- Rudall River National Park, in the southwestern section of the Great Sandy Desert, is the largest national park on the Australian continent. It is essentially a conservation area with a few Aboriginal communities and not much else, other than some newly discovered and extensive uranium deposits.

- The sand dunes of the Gibson Desert are fixed. The Great Victoria Desert, however, has sand dunes that are actively developing, or moving. This means that the ridges are constantly molded by the wind and change with time.

- Hurricanes called "willy-willies" hit Australia's west coast between Broome and Onslow and tear southeast across the Great Sandy Desert, bringing major rain and floods to the area.

- The Western Australian Desert has an average elevation of 400–600 meters (1,300–3,000 ft.) above sea level. On top of this fairly level landscape lie large plateaus, such as the Kimberley, that rise to about 1,500 meters (5,000 ft.).

- A huge amount of diamonds were discovered in the Kimberley region in 1979. Within 10 years Australia had become the world's largest diamond exporter. The country now accounts for 30 percent of diamond production worldwide.

- The average rainfall in the Western Australian Desert is less than 25 centimeters (10 in.) per year. Generally, rainfall is variable, with long droughts followed by rains and flooding.

- The type of vegetation found in the Western Australian Desert includes semiarid savannas, hummock grasslands, tussock grasslands, and shrub steppes.

- A number of Aboriginal reserves are dotted throughout the Western Australian Desert, many of which still maintain their traditional ways within tribal societies.

- The animals of the Western Australian Desert must be able to survive under extremely dry conditions and unpredictable rainfall. They are either nomadic and able to go wherever water can be found, or have a special feature that helps them cope with the extreme heat and scarcity of water. The water-holding frog *Cyclorama*, for example, burrows deep into the sand, and its outer skin forms a protective sac around it that it can then fill with water.

# Central Australian Desert

**1. Great Victoria Desert**
This sandy wasteland in Western and Southern Australia stretches between the Gibson Desert in the north and the Nullarbor Plain in the south, and from the town of Kalgoorlie in the west almost to the Stuart Range in the east. There are no paved highways running through this desolate region. In the eastern part are important Aboriginal lands.

**2. Macdonnell Ranges**
This system of bare sandstone and quartzite mountain ranges rises at the heart of Australia's arid inland plateau and features striking rock domes, including Uluru. The ranges' streams and rivers make the surrounding region the best-watered in Central Australia.

**3. Mount Zeil**
Mount Zeil, in the Macdonnell Ranges near Alice Springs, is the highest point in the Central Australian Desert. It rises to a height of 1,511 meters (4,957 ft.).

**4. Alice Springs**
For 10,000 years the site of this remote town was an Aboriginal stopover on the way to the waterholes in the Macdonnell Ranges. It became a permanent settlement only in the 1870s, when Charles Todd, the then Superintendent of Telegraphs, established an overland telegraph station here, naming it after his wife, Alice.

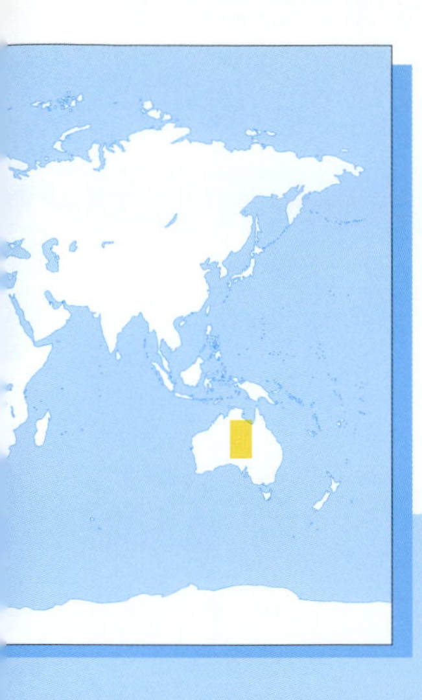

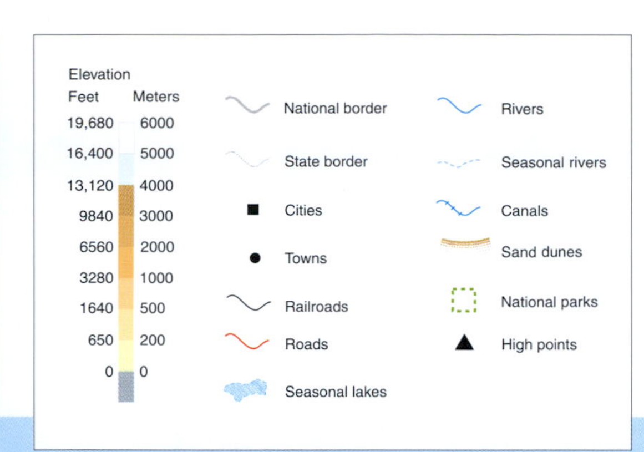


### 5. Uluru (Ayers Rock)

This vast oval-shaped sandstone rock, today officially known by its Aboriginal name, Uluru, but formerly called Ayers Rock, is located in the middle of the Central Australian Desert and is the most visited natural site in Australia. The area surrounding the rock became an Aboriginal reserve in 1920 and a national park in 1958. In 1977 the park was proclaimed a UNESCO World Heritage site.

### 6. Mount Olga

At 1,070 meters (3,507 ft.), this is the highest of a roughly circular group of more than 30 rock domes, collectively called the Olgas, that rise from the desert plains to the west of Uluru. The Aboriginal name for the Olgas is Katajuta, meaning "many heads."

### 7. Stuart Highway

The Stuart Highway runs roughly north–south for 2,700 kilometers (1,678 mi.), from Darwin on the Timor Sea to Adelaide on the South Australian coast, passing through the Central Australian Desert. For strategic reasons the road was upgraded during World War II and is now entirely paved. The highway is named after John McDouall Stuart, who explored the surrounding region in the 1860s.

### 8. Simpson Desert

Covering some 145,000 square kilometers (56,000 sq. mi.), this desert region is almost entirely uninhabited except by a small number of Aborigines. The desert features parallel sand ridges, about 20–37 meters (70–120 ft.) high, that run northwest to southeast for distances of up to 160 kilometers (100 mi.). There is little vegetation other than spinifex grass, which grows between the dunes, and the only water is found in rock holes and dry river soaks. Nevertheless, the desert is a haven for some of Australia's rarest desert animals, including the fat-tailed marsupial mouse.

### 9. Ghan railroad

This railroad, running from Alice Springs in the Northern Territory to the South Australian capital, Adelaide, is the only one in the Central Australian Desert. It was named in honor of the Afghan camel drivers who transported passengers, goods, and mail through the desert before the advent of the railroad.

### 10. Nickel deposits

Large deposits of nickel exist in the Musgrave mountain ranges on the border between South Australia and the Northern Territory.

### 11. Lake Eyre

Covering 9,324 square kilometers (3,600 sq. mi.), Lake Eyre is the largest salt lake in Australia (*see illus.*, p. 162). Much of the lake is dry, except when rare floods fill up the basin. The Lake Eyre area has the least amount of rainfall in Australia, averaging as little as 13 centimeters (5 in.) a year. The lake is named after the explorer Edward John Eyre.

### 12. Coober Pedy

The town of Coober Pedy is one of the major centers in Australia for the mining of white opal (*see* p. 177 *and illus.*). Opal was first discovered here in 1915, but the town burgeoned only at the end of World War II, when returning soldiers were drawn here to try their luck as prospectors. In January and February the town is besieged by dust storms that turn the sky a bright orange.

### 13. Mount Isa

This Queensland settlement developed as a center for the region's rich mineral resources, which include lead, zinc, silver, and copper. These were originally discovered in 1923 by John Miles, who established the Mount Isa Mines a year later. The town now accounts for two-thirds of Australia's copper output.

## Fact File

- The first base of the Royal Flying Doctors, founded in order to bring medical help to settlers in the remote outback, can be seen at Mount Isa.
- About one-third of the Central Australian Desert now belongs to the Aborigines and can be visited only by those who hold a permit or an invitation.
- The last big gold rush in Australia, in 1933, was centered around Tennant Creek – a town in the middle of the Northern Territory on the Stuart Highway north of Alice Springs. Copper was later discovered in the area in the 1950s.
- About 100 kilometers (62 mi.) south of Tennant Creek is an unusual geographical phenomenon called the Devil's Marbles. It is an area strewn with big granite boulders, which the Aborigines believe are the eggs of the Rainbow Serpent spirit.
- Australia's first international airport was located in the Central Australian Desert at Daly Waters, about 400 kilometers (250 mi.) north of Tennant Creek. It was used as a refueling stop by Quantas airlines in the 1930s on the Sydney to London route, and later by bombers in World War II.
- In the 1970s the government granted land rights to many Aborigines. This resulted in a number of small Aboriginal communities of fewer than 100 people, known as outstations, scattered throughout the desert regions. These communities function without many basic facilities such as electricity.
- The Tanami Desert in the Northern Territory is almost entirely owned by the Aborigines. An important Aboriginal event called the Yuendumu Sports and Cultural Festival is held here in the Yuendumu community, 290 kilometers (180 mi.) northwest of Alice Springs, for a weekend in early August. Visitors are welcome and, exceptionally, no permits are needed.
- The Central Australian Desert is the heart of Australia's cattle industry. Cattle stations have been here since 1880. Today stations are often owned by big cattle companies and cover huge areas up to 5,180 square kilometers (2,000 sq. mi.), with some 20,000 cattle managed by only about 15 people. Australia is the largest exporter of beef in the world.
- Finke Gorge National Park, between Alice Springs and Uluru, is an oasis of vegetation amid the surrounding desert, nourished by the mainly underground Finke River. Some of the species found here, such as the cabbage palms that grow 20 meters (66 ft.) high, exist nowhere else in the country.

## The Last Frontier

The Australian desert has been called "the Last Frontier," since it was one of the last deserts to be officially discovered and explored by Europeans. Australians themselves often refer to it as "the outback." At least one-third of the Australian continent, in the interior of the country, is covered by desert. The lack of cloud cover in this inland region frequently causes temperatures to climb to over 38°C (100°F).

### Geography of the Region

The terrain is generally characterized by arid lowland plains, broken up by areas of spectacular isolated granite rocks, called inselbergs, which rise above the flat plains, and by sand dunes. There are also upland regions of elevated sandstone plateaus. Although the average rainfall in the desert areas amounts to a mere 250 millimeters (10 in.) a year, when rain does fall, the effect is dramatic.

**An emu (*Dromaius novaehollandiae*) speeds across the Australian desert. Despite being the second-largest bird on earth, the emu is able to sprint at 60 kilometers per hour (37 mph).**

**Lake Eyre in the Central Australian Desert is the largest salt lake in Australia. It also contains the lowest point in Australia – 16 meters (52 ft.) below sea level.**

**The desert settlement of Alice Springs, Northern Territory, stands virtually at the center of the Australian continent in the Macdonnell Ranges and is an important mining and pastoral center.**

This is because the soil in the higher plateau regions is often made of layers of impermeable rock with only very little vegetation, so the runoff from the rainfall is usually very rapid.

The fast-running streams and rivers erode gullies and carry large amounts of sediment down from the plateaus to the lower plains, including the sand that makes up the desert dunes. A number of rivers drain into Lake Eyre, which, at 16 meters (52 ft.) below sea level, is the lowest area of the desert and of the whole of Australia. Lake Eyre is generally a dry salt lake in most parts, but about twice a century flood waters fill up the lake's basin.

### Animal and Plant Life

The main types of vegetation in the Australian desert are species of *Acacia, Eremophilia,* and *Casuarina*. Some eucalyptus species have also adapted themselves to the climate. Animals who can survive in these arid conditions include the marsupial mole, the water-holding frog, and the budgerigar. Other birds found in the desert include the Australian magpie and the galah – a species of cockatoo.

## PEOPLE OF THE DESERT

Most of the inhabitants of the Australian desert are Aborigines. They are thought to have arrived from Asia about 40,000 years ago. Nowadays, owing to an Aboriginal campaign for land rights in the 1980s, areas of the desert have been granted to various Aboriginal groups. The Aborigines are traditionally nomadic hunter-gatherers. They are extremely religious and their beliefs are based on a very close relationship with nature. They form social groups called "estate groups" that share responsibility for specific sites and territories, known as their "estate." Estates are attributed to the group through "Dreaming" – a spiritual state outside space and time that represents the continuity of life, and in which knowledge is imparted by the spirits to adult males in the group, who also act as guardians of the sacred sites and who create rituals to replenish and bless the land. Ownership of these territories can not be transferred, which is why the land rights campaign was so important to the Aborigines. Traditionally, most of the group did not live on the estate but in "bands" that stretched across it and included areas where they would go to find food and water. In desert regions, however, the boundaries of these bands were more flexible, since droughts would often force groups off of their own bands and into those of other groups. Aborigines currently make up about 2 percent of Australia's population.

# The Arctic

## 1. The Arctic Circle

The Arctic Circle is a line, or parallel, of latitude around the earth located at about 66°30' north. It marks the southern boundary of the area within which, for one day or more each year, the sun never sets (on or about June 21) or rises (on or about December 21). The length of the continuous day or night increases northward, ranging from one day near the Arctic Circle boundary to six months at the north pole.

## 2. The Arctic Ocean

Most of the Arctic consists not of land but of sea. The Arctic Ocean covers an area of 14.1 million square kilometers (5.4 million sq. mi.), with the north pole at its approximate center. The world's smallest ocean – only about one-sixth the size of the next largest, the Indian Ocean – the Arctic has a maximum depth of 5,502 meters (18,050 ft.). The ocean is mostly covered by ice, which is permanent above 75° north and seasonal between 60° and 75° north. It is divided into two major basins – the Amerasia and the Eurasia – by the Lomonosov Ridge. The main circulation point between the Arctic and other oceans is via the Fram Strait between Svalbard and Greenland.

## 3. Glaciation

Less than two-fifths of land in the Arctic is permanently covered with ice and can be

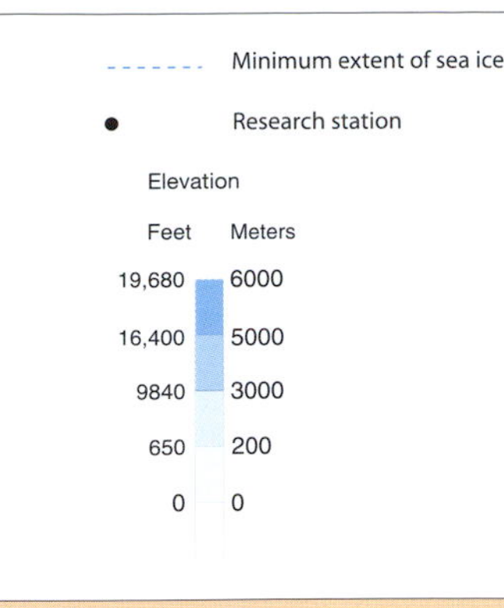


called polar desert. In most Arctic regions temperatures are relatively warm or there is scant snowfall, and so plant and animal life are relatively rich. The most important polar desert in the Arctic is situated in Greenland.

**4. Greenland Glacier**
The Greenland Glacier is the Arctic's largest glacier, and one of the few true polar deserts in the northern hemisphere. It covers 1.7 million square kilometers (650,000 sq. mi.) – almost 80 percent of Greenland – and is 2,527 kilometers (1,570 mi.) long by 966 kilometers (600 mi.) wide. The glacier lies in a shallow, bowl-like depression. In its center the ice extends down to more than 300 meters (1,000 ft.) below sea level.

**5. The coldest place**
The lowest land temperature ever recorded in the Arctic was in northern Greenland – a chilling -70°C (-94°F).

**6. Vatna Glacier**
This glacier in southeastern Iceland covers more than 8,400 square kilometers (3,200 sq. mi.) and is sustained by the island's heavy snowfalls. The glacier conceals numerous active volcanoes, including Grimsvötn, which erupts every 6 to 10 years, sometimes causing massive floods called *jökulhlaup* ("glacier runs").

**7. Svalbard**
Some 60 percent of Svalbard ("Cold Coast") is covered with ice or snowfields.This Arctic archipelago nevertheless supports a human population of about 3,000, most of whom live in the capital, Longyearbyen, on the island of Spitzbergen. Since 1920 the islands have been part of Norway, but other countries, notably Russia, have retained coal-mining rights to the land. Animal life on the islands includes polar bear, reindeer, and arctic fox, as well as marine fauna. The islands have often served as a base for polar scientific expeditions.

**8. Franz Josef Land**
This desolate archipelago of 191 islands is 85-percent covered with ice. Discovered by an Austrian and named in honor of the Austrian emperor in 1873, the islands were annexed by Russia in 1926 and today constitute that country's most northerly territory. The islands' climate is very severe, with average winter temperatures as low as -22ºC (-8ºF) and average summer temperatures reaching barely 2ºC (35ºF). Vegetation consists mainly of lichens, mosses, and almost 40 species of Arctic flowering plants. Animals include polar bears and arctic foxes. Human habitation is confined to a few permanent weather stations.

**9. Three great rivers**
The Russian area of the Arctic includes three of the world's longest rivers – the Ob', the Yenisey, and the Lena. All of these flow northward and empty into the Arctic Sea, but in their northern reaches are ice-free for a only a few months each summer.

**10. Bering Strait**
Russia was the first nation to explore the Arctic. In 1733 the czar sent an expedition to Siberia under the command of the Danish navigator Vitus Bering (1681–1741) to determine the existence of a northern sea passage between America and Asia. The expedition resulted in the discovery of the 85-kilometer-wide (53 mi.) strait between Alaska and Siberia that was later named after Bering.

**11. Oil resources**
Two of the world's largest areas for oil and natural-gas resources are located in the Arctic. Alaska's Arctic oil fields produce one-fifth of the United States' output, while the oil and natural-gas fields of northwestern Siberia, which cover an area of some 971,246 square kilometers (375,000 sq. mi.), account for almost all of Russia's output.

**12. Diamonds**
The Arctic's largest deposits of diamonds can be found in the Sakha region in Russia.

# Fact File

- Parts of the Arctic constitute the northernmost desert regions on earth.
- Only two-fifths of the Arctic's land surface is covered by permanent ice.
- The Arctic includes parts of eight countries: Canada, Finland, Greenland, Iceland, Norway, Russia, Sweden, and the United States.
- The first people to inhabit the Arctic were the Paleo-Siberians ("Old Siberians"), who settled in northeastern Siberia's unglaciated lowlands in 5000 B.C.
- The average life expectancy for people living in the Arctic ranges from 62 years in Greenland to 75 years in Norway.
- The Soviet Union was the first nation to set up a scientific station in the Arctic. It achieved this by landing an airplane equipped with skis on the Arctic ice.
- Raging cyclones often tear across the Arctic. This is because the lack of trees and the flat, smooth surface of ice over the Arctic Ocean offer no friction or resistance against the wind.
- The word "arctic" comes from the Greek *arktos*, meaning "bear." It originated from the northern constellation of the Bear, which was used to designate the area inside the Arctic Circle.
- The most widely found mammals of the Arctic are the polar bear, caribou, musk ox, arctic wolf, arctic fox, arctic weasel, arctic hare, and two species of lemming.
- North of the timberline, Russia and North America are covered almost entirely by permafrost. Permafrost is permanently frozen earth where the temperature is always below 0°C (32°F). Near the northern coasts of Russia and North America, the permafrost extends down as far as 450–600 meters (1,500–2,000 ft.), with ice making up almost four-fifths of its volume.
- Arctic mammals are unable to hibernate because of the cold and therefore stay active all year round.
- Submerged glacial valleys, known as fjords, are an important feature of the Arctic and can be found in Greenland, eastern and western areas of Iceland, the Norwegian coast, the east coast of Canada, and the southern coast of Alaska.

# The Antarctic

**1. The Antarctic Circle**
The Antarctic Circle is the line of latitude around the earth at 66°30' south within which the sun does not set (on or around December 22) or rise (on or around June 21) for one day or more per year. The days and nights are longer in areas that are closer to the south pole and vary from one day to six months. This phenomenon is caused by the earth's axis being tilted by 23.5° to the prime vertical, instead of being perpendicular to the equator.

**2. Ice sheets**
About 10 percent of Antarctica's ice is made up of ice sheets, or shelves, which bridge the continent's bays and seas. The main ones are the Filchner Ice Shelf, the Ross Ice Shelf, and the Ronne Ice Shelf.

**3. Lowest temperature**
Antarctica is the world's coldest continent, with winter temperatures averaging about -25°C (-9°F) on the coast and -55°C (-27°F) in the interior. The lowest temperature ever recorded was -89.2°C (-128.6° F), which was at Russia's Vostok Station on the inland ice sheet. The coldest month is August, just before the return of day.

**4. Earthquakes**
Only one significant earthquake has ever been recorded in

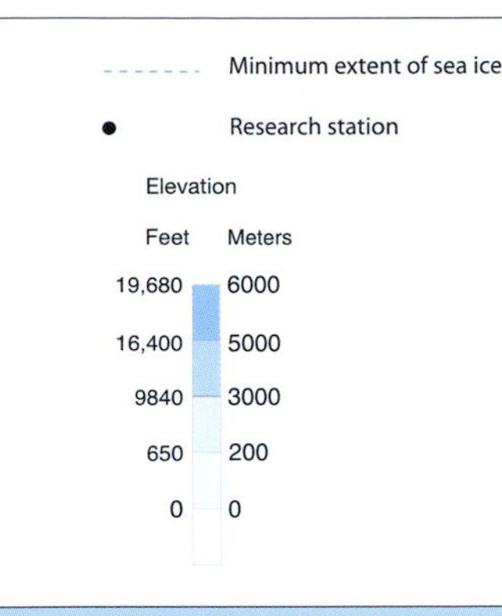


Antarctica. This was in the Bellingshausen Sea in 1977, registering 6.4 on the Richter scale. The sea is named after one of the first discovers of Antarctica, Fabian Gottlieb von Bellingshausen, who led a Russian expedition around the continent in 1819–1821.

**5. Largest scientific station**
McMurdo Station, at the tip of Ross Island, is Antarctica's largest scientific station and Is run by the United States.

**6. Ross Sea and Ice Shelf**
Ross Sea and the Ross Ice Shelf are named after the British explorer James Clark Ross, who discovered them on an expedition in 1839–1843.

**7. Subtropical Convergence**
The southern regions of the Atlantic, Pacific, and Indian oceans are known as the Southern Ocean, which contains about 10 percent of the earth's water. Its northern boundary is marked by the Subtropical Convergence at a latitude of 40° south. This is where warm subtropical surface currents mix with cold Antarctic waters.

**8. South pole**
The south pole marks the southernmost point of the earth. The site was first reached by Norwegian explorer Roald Amundsen in December 1911 and by British explorer Robert F. Scott a month later in 1912.

**9. The ozone hole**
A hole in the ozone layer was discovered over East Antarctica in 1977, but was made public only in 1985. The hole is located around an area called the polar vortex, where air is caught in a circulation pattern brought on by the end of the long Antarctic winter. Scientists believe the loss of ozone may be due to complex chemical reactions in the atmosphere to the sun's return. The ozone hole is now being monitored by NASA using high-flying aircraft.

**10. Dinosaur fossils**
Dinosaur fossils from the Jurassic period (208–146 million years ago) were discovered in the Transantarctic Mountains near the south pole. This suggests that the continent's climate was a great deal milder at that time.

**11. Wind speeds**
The highest wind speeds ever recorded in Antarctica were at Australia's Mawson Station on MacRobertson Land, with speeds as high as 237 kilometers per hour (147 mph). Most coastal regions of East Antarctica suffer from katabatic winds – dense, cold winds that flow down the slopes from the high interior. The wind is so strong it picks up snow from the ground and whirls it through the air, creating a blizzard effect. A fierce west wind rips around the entire Antarctic continent in a clockwise direction, dragging the currents along beneath it.

**12. The highest point**
The highest point in Antarctica is the Vinson Massif in the Sentinel Range, rising to 4,897 meters (16,864 ft.).

**13. Volcanoes**
Active volcanoes exist in Ellsworth Land and Marie Byrd Land and on the coasts of Victoria Land and the Antarctic Peninsula. East Antarctica has one volcano – Gaussberg on the coast at 90° east. Mount Erebus, on Ross Island, has shown increased activity since the mid-1970s.

**14. Dumont D'Urville Sea**
This sea in the Southern Ocean is named after the man who led an Antarctic expedition in 1837–1840 and discovered Adélie Land, which he claimed for France.

**15. Meteorites**
A large number of meteorites have been found in Antarctica, mostly near the Transantarctic Mountains and the Japanese Syowa Station. The majority are thought to be from asteroids or, less frequently, comets. A piece of the moon was found in 1982, and in 1984 a meteorite from Mars was discovered that some scientists believe may contain fossils of Martian life.

# Fact File

- Antarctica, which means "opposite to the Arctic," is the earth's fifth-largest continent, covering an area of 14.2 million square kilometers (5.5 million sq. mi.).

- An ice sheet about 2,000 meters (6,500 ft.) thick covers 30 million cubic kilometers (7 million cu. mi.) of the continent of Antarctica and makes up about 90 percent of the world's ice.

- Antarctica is divided into two subcontinents: East Antarctica and West Antarctica. East Antarctica – the larger area – is a high, ice-covered plateau, while West Antarctica is a string of mountainous islands covered and bound together by ice.

- The Antarctic Treaty, signed in 1959 by 12 nations, preserves the continent of Antarctica for nonmilitary scientific studies. The treaty was proposed during the International Geophysical Year (IGY) in 1957–1958 and is still considered a model of international diplomacy.

- Antarctica is the world's highest continent. Its average elevation is 2,286 meters (7,500 ft.) above sea level. The next highest continent, Asia, averages only 914 meters (3,000 ft.).

- Only a few plant species can survive the cold, dark Antarctic winters. These include bacteria, lichens, freshwater algae, liverworts, molds, mosses, yeasts, and other fungi.

- If all the ice in Antarctica were to suddenly melt, sea levels around the world would tend to rise by 45 to 60 meters (150–200 ft.).

- All of Antarctica's mammals are marine animals. There are a number of seal species – such as the crabeater, elephant, fur, leopard, and Weddell seals – as well as porpoises, dolphins, and whales. The main species of whale are the killer whale, sperm whale, and the bottle-nosed whale.

- Seven countries claimed sovereignty over portions of Antarctica between 1908 and 1942. They were Argentina, Australia, Britain, Chile, France, New Zealand, and Norway.

- The average precipitation on Antarctica's inland polar plateau is 6 centimeters (2 in.) per year, whereas the coastal average is about 50 centimeters (20 in.). Rainfall is very rare and most water falls in the form of snow.

- The first person to fly over the south pole was U.S. naval officer Richard E. Byrd in 1929.

# WEALTH FROM THE DESERT

*Deserts are treasure-houses of some of the earth's most sought-after resources, including oil. Exploitation of such resources has often been bought at great environmental cost, however, and humankind is only now realizing the true value of the desert – as a precious natural habitat.*

Since ancient times deserts have been used as a source of metals and minerals. In ancient Egypt, for instance, gold was regularly extracted from the Western Desert and used lavishly in the creation of royal artefacts, such as those that adorned the tomb of Tutankhamen. The irrigation of semiarid desert fringes, too, is an age-old practice and was the basis of some of the world's oldest civilizations (*see* pp. 132–133). However, it is only in the last hundred years or so that the development of new technologies and transportation infrastructures has enabled the wholesale exploitation of desert wealth, with the result that pristine wildernesses have been transformed into semi-industrialized landscapes of mines, oil wells, highways, and pipelines. Above all, it was the discovery of oil, especially in the Middle East, that began a new age in the history of the desert.

**Trucks are dwarfed by the vast scale of the copper mine at Chuquicamata in Chile's Atacama Desert. The open-pit mine is now more than 3 kilometers (2 mi.) long and is the largest copper mine in the world.**

Exploitation of desert areas has often been reckless. Because deserts are often perceived as "empty," any degree of human intervention in the landscape has appeared justifiable. Mining, drilling, and irrigation have caused severe environmental damage in many arid areas. Moreover, as deserts have become sites of economic exploitation, they have also become subjects of political and military dispute. The Gulf War of the early 1990s was to a large degree an "oil war," fought on all sides to preserve access and rights to what is an increasingly precious commodity.

As our knowledge of deserts has improved – both in terms of their place in the global environmental jigsaw and their value as ecological havens – a more balanced approach to their exploitation has begun to emerge. The development of solar-energy plants offers a model for such an approach, in which a renewable desert resource – in this case, sunlight – is exploited at minimal environmental cost. The preservation of Antarctica – known to be a potentially rich source of oil – under international law is another promising sign of changing attitudes to the desert.

**In the late 20th century the nations of the Middle East grew rich on the oil deposits that lie beneath the desert surface. Saudi Arabia is one of the world's biggest oil producers, in 2002 generating over eight million barrels per day.**

## Petroleum

The pressing demand for crude oil, or petroleum, was a powerful factor in the shaping of late-20th-century history and, as resources diminish, is likely to play an ever more crucial, and indeed fractious, role in 21st-century global politics. Crude oil is readily portable and, in refined states, furnishes not only major fuels, such as gasoline, kerosene, and heating oils for electricity-generating power plants, but also a vast array of other products, including plastics, artificial fibers and textiles, pigments and bases for paints, lubricating oils, and bitumens, tars, and asphalts.

Almost two-thirds of global reserves of crude oil are sited under the deserts of the Middle East and the arid fringes of North Africa and western Asia. There are also major oil fields in nondesert regions, including West Africa, the North Sea, the Gulf of Mexico, Alaska, and western Siberia. Oil provides some 40 percent of the world's commercial energy needs. However, this figure is a global average. In largely rural nations such as China only 17 percent of commercial energy needs are met by oil. In industrialized Canada, where there is an abundance of hydroelectric power, the figure is about 35 percent. In many other industrialized countries, however, the proportion exceeds 70 percent. Some African countries, such as Burkina Faso, Senegal, Gambia, and Sierra Leone, are more than 95-percent dependent on oil for their energy needs.

With their lack of soil and vegetation cover, desert regions are relatively easy places to search for oil. Surveying methods include infrared and other kinds of imaging from satellites and airplanes and the use of small explosions or thumper machines to send sonic vibrations and seismic waves through the ground, whose progress is then analyzed by computer. If an oil or gas field seems likely, test boreholes are drilled at various sites in the area to check the size and depth of the field before production starts. Once the bores are drilled, tapped, and producing, and the pipelines are laid to ports or refineries, operation of an oil field is relatively automatic. Few personnel have to endure the harsh desert climate and conditions.

### Diminishing reserves

The huge importance of oil in today's industry, manufacturing, and transportation means that it plays a central role in both the global economy and international politics. Some oil-rich nations of the Middle East have become hugely wealthy and exert great leverage over international affairs. The threat of dwindling oil reserves has been of major political concern for several decades. In the early 1970s some experts estimated that world energy reserves, including oil, would last only another 40 to 50 years. However, at the turn of the millennium and despite rocketing oil consumption, particularly in the fast-developing "tiger economies" of East Asia, similar

## WHO HAS THE OIL?

This table shows how the world's recoverable oil reserves are shared among various nations.

|  |  |
|---|---|
| Saudi Arabia | 25% |
| Iraq | 10% |
| Iran | 9% |
| United Arab Emirates | 9% |
| Kuwait | 9% |
| Venezuela | 7% |
| Mexico | 5% |
| Russia | 5% |
| Libya | 4% |
| United States | 3% |
| Others | 14% |

predictions still held sway. Between 1973 and 1993 proven recoverable reserves of petroleum rose by 60 percent, and those of gas by more than 140 percent. This was partly due to major reevaluations of the Middle Eastern oil fields in the late 1980s.

Since 1990 finds of petroleum have roughly matched increasing world consumption. Nevertheless, reserves in non-Middle Eastern oil fields have fallen – down 10 percent in Russia and its associated states and 14 percent in the United States from 1984 to 1994 – so the desert nations of the Middle East are likely to exert greater influence in the future.

**Oil wells blaze and an oil slick forms in the Kuwaiti desert after the Gulf War in 1991. Retreating Iraqi forces set light to more than 600 oil wells, turning the Kuwaiti desert into a polluted wasteland. The oil fires burned for a year and a half, and some 60 million barrels of oil gushed out of the broken wellheads onto the land, contaminating freshwater aquifers.**

Over the longer term, however, oil will certainly run out at current rates of use. Estimates vary from the pessimistic 50 years predicted by many green and conservation movements to the more optimistic 100 years or longer suggested by supporters of the petroleum industry.

### Oil wars

Crude oil is so valuable that it is often called "black gold." Wars have been fought for its possession, and shortages of petroleum products such as gasoline can trigger strikes and revolutions. In 1908 fields were found in the arid regions of Persia (now Iran), followed in the late 1930s by discoveries of massive fields in Saudi Arabia. In 1973 and again in 1979 some Middle East countries raised oil prices, and the industrialized West was forced to acknowledge its dependence on crude petroleum and its products as well as on the goodwill of the Arab nations. Worldwide recessions followed each time.

The Gulf War of 1990 to 1991 is a warning example of the kind of conflict that this precious commodity can cause. On August 2, 1990 the army of Saddam Hussein (1937–), president of Iraq, invaded and occupied neighboring Kuwait. One of the major reasons proffered by Hussein's regime for its action was that Kuwait was overproducing oil and selling too much too cheaply, thereby driving down prices and causing economic hardship to other petroleum exporters. Many commentators, however, suggested that Iraq's real motive was to annex Kuwait's rich crude-oil resources for itself.

Most of the international community supported the United Nations' request for Iraq to withdraw from Kuwait by January 15, 1991. This did not occur, and the next day air strikes against Iraq marked the beginning of Operation Desert Storm and the Gulf Conflict. The military campaign was led by the United States and other Western nations, whom some observers saw as less interested in the rights of the Kuwaiti people than in the threat to oil supplies if Iraq were to became too powerful in the Middle East.

The sandy deserts of the Middle East, with no visible wealth to the casual observer, became a fiercely contested prize. By February 27 the conflict was largely at an end, Iraq was defeated, and a United Nations-brokered condition imposed that some 30 percent of Iraq's future revenue from selling petroleum should be put into a fund for helping the conflict's victims. Massive acts of ecoterrorism took place as the retreating Iraqi forces deliberately set fire to oil wells and spilled millions of liters of crude oil into the waters of the Gulf.

## Metals and Minerals

The prevalence of petroleum under land that is now desert is an accident of prehistory. Much the same is true of the inorganic geological processes that have led to certain metals and minerals being concentrated in rocks of desert regions. Mining and quarrying companies face numerous problems working in hot, dry conditions. Lack of local habitation resources and infrastructure compared to those of more equable, well-populated regions means that almost every necessity and comfort, including fuel, food, and water itself, must be transported – usually at high cost. However, the remote, barren nature of desert regions means that there are usually few people to protest about habitat disturbance or environmental degradation.

### Fertilizer from the desert

Like petroleum, the mineral wealth found in some desert areas has proved a source of conflict. Many peripheral arid areas have been ignored until their hidden wealth is revealed, at which point the land becomes an object of contention for competing neighboring nations. Rich deposits of phosphates and nitrates found in and around the Atacama Desert in South America created just such regional rivalries.

The Atacama's phosphates and nitrates formed relatively recently, as minerals became concentrated in low-lying basins after successive rare rains and floods. Harvesting these dry, salty basins became very profitable during the second half of the 19th century, as deposits were extracted and processed to supply the burgeoning Western chemical industry and also the beginnings of a world boom in intensive agriculture. The Atacama's nitrate beds were the only major source of inorganic nitrogen fertilizers. Such a valuable resource did not go uncontested, however.

In the period after 1850 the Atacama Desert lay in Bolivian territory, but it was Bolivia's neighbor Chile that set up the mines to exploit the desert's nitrates. In 1878 Bolivia began to levy charges for the privilege, a move that was deeply resented by the Chileans. As the dispute flared, the Bolivians took military control of much of the Atacama area, including its principal port and nitrate outlet, Antofagasta. Chile fought back, not only occupying the port but invading Bolivia's ally, Peru. The War of the Pacific, as the dispute is known, only ended in 1884 after Bolivia agreed to relinquish its only coastal province.

As well as being excellent fertilizers, nitrates are also an important ingredient in the production of gunpowder. During World War I (1914–1918) the Allies blockaded Germany's access to the Atacama's nitrate fields in order to reduce its ability to manufacture this explosive mixture. This spurred German chemists Fritz Haber (1868–1934) and Carl Bosch (1874–1940) to develop a synthetic means of producing ammonia, a compound of nitrogen and hydrogen that can be used to make both explosives and fertilizers. As the Haber–Bosch process became generally available, the Atacama's nitrate deposits lost their earlier importance, and production faded after about 1920.

**Villagers extract salt from pits dug into the groundwater at Bilma Oasis in Saharan Niger. Small-scale extraction of minerals and metals such as salt, gold, silver, and copper has taken place in deserts for thousands of years.**

## Greening the Desert

From earliest times there have been attempts to irrigate and fertilize the desert, with varying degrees of success. Attempts continue today, as population growth puts pressure on many countries to use as much space as possible for growing food. Projects to overcome deserts' critical absences of moisture and nutrients while making the best of the abundant solar heat and light are under development around the world.

Some of these projects involve relatively natural, that is to to say, traditional, methods of farming – using dry-adapted crops, such as jojoba seeds, morama beans, yeheb, and guayule, or raising drought-coping livestock such as antelope, goats, kangaroo, emu, and ostrich. However, true deserts lack the rainfall and soil to make such projects sustainable in the long term, and most of these so-called desert farms, both arable and pastoral, are generally confined to semiarid desert fringes.

Traditional methods of irrigation may help to sustain some desert farms in the medium term. Underground irrigation tunnels called qanats have been cut into rocks in the Middle East – especially in Persia – for thousands of years. These tap groundwater from aquifers at mother wells and channel it out and up to the surface for crop growth or watering livestock, as well as supplying human

**Large-scale irrigation of many semiarid and arid areas of Israel has created extensive croplands. Despite its largely desert terrain the country has become an important exporter of citrus fruits and vegetables.**

demand. The construction of qanats is laborious – dug by hand, the tunnels grow at rate of about 1 kilometer (3,280 ft.) per year, with regular vertical shafts to the surface every 30 to 50 meters (100–165 ft.), for ventilation and access.

There are an estimated 40,000 qanats in Iran, some up to about 40 kilometers (25 mi.) long. They are also used by rural farms and villages in arid Egypt, Yemen, Iraq, Afghanistan, Pakistan, and China, as well as in Algeria, Morocco, and Spain. In Algeria, where they are known as *foggara*s, they help to irrigate about one-half of all croplands. In the Americas the great Chimú Empire, which reached its apogee in the 13th century in the Sechura Desert, used qanats to bring water from the Andes foothills that lay to the east.

## Hi-tech developments

Qanats tap groundwater up to its natural level of replenishment and no farther, and are therefore in the main self-regulating and sustainable. Modern intensive agriculture demands more copious, and more reliable, supplies of water. This has led to several experiments in high-technology desert farms where pumps suck water from boreholes to supply intensive crops. Such techniques have notably been used in Israel's Negev Desert, where farms produce fruit and vegetables mainly for export.

However successful in the short term, such plans to make the desert "bloom" are in the long term unsustainable – a mirage as intoxicating and useless as any phantom oasis. Projects like that in the Negev use water at a very high rate, and supplies cannot be replenished naturally. Many projects also require huge capital investment, involve high running costs, and are based on complex, sophisticated machinery that cannot be maintained or repaired using local skills and materials. Although some estimates suggest that groundwater exploited for such intensive agriculture will last for as long as 200 years, such supplies will inevitably become more saline as salty minerals are sucked up from deeper in the rocks.

Updated versions of traditional arid-land grazing – using satellite imaging to find water – common in the Middle East, may be less environmentally damaging. However, such techniques can encourage overgrazing and the degradation of already vulnerable marginal lands.

## Future technologies

The development of genetically modified (GM) crops may hold better promise for a blooming and fruitful desert. In strong sunlight traditional crop plants close the tiny pores called stomata in their leaves to prevent too much water loss from their tissues by transpiration. This has the effect of reducing the incoming supply of carbon dioxide from the atmosphere, which in turn slows the vital process of photosynthesis. Prolonged strong sunlight also causes various forms of "active oxygen," such as hydrogen peroxide, to build up in plant cells, which can be potentially damaging.

Research is under way to identify plant genes for various enzyme substances that encourage the biochemical change of active oxygen into water. Early trials show that altered plants are less likely to wither and die from too much sun. Another approach is to manipulate the plant enzyme RuBisCo – based on a chemical called ribulose biphosphate – which controls the rate at which atmospheric carbon dioxide is incorporated by photosynthesis into plant tissues. The incorporation of altered RuBisCo genes into crop plants would increase the efficiency of carbon-dioxide usage. Several similar research projects are underway, involving a mix of genetic modification and traditional breeding techniques, with the focus on drought-resistant native plant species. These could well be the best way of making the semidesert environment yield useful crops over the long term.

### Energy from the Desert

Heat and light are both forms of solar energy, and both are plentiful in most desert regions. Large solar power plants have been established in the American Southwest, the Sahara, southern Africa, and Australia. The cost of trapping the sun's energy has fallen by more than half between 1885 and 2000, due mainly to increased research into, and mass production of, the photovoltaic cells that convert light energy into electrical energy. Arrays of these cells are combined into large solar panels, which, although costly to manufacture, are virtually maintenance-free once installed.

Sunlight is a dilute form of energy and thousands of hectares of solar panels are needed to generate industrial quantities of electricity. In many ways deserts, with their lack of cloud cover, long hours of exposure to bright sun, and mainly low-profile wildlife, are ideal sites for solar energy farms. In the deserts of Israel, Saudi Arabia, Australia, and elsewhere, small-scale domestic solar panels already power radios, televisions, and similar low-wattage electrical equipment.

In addition to light, the sun also radiates heat. A power plant called a solar furnace has been developed to collect this infrared radiation, usually by means of mirrors or similar reflectors that direct the rays onto a central collector. One of the world's first solar furnaces, called Solar One, near Barstow in the Mojave Desert, had 1,818 such heat-focusing mirrors, or heliostats, and began operating in 1982. During the six years of its operation, Solar One produced 38,000 megawatt-hours. Between 1996 and 1999 Solar Two, at the same site, piloted a more efficient method of storing the energy collected by the heliostats, involving the use of molten-salt power towers.

A new proposal for solar power in the desert is termed the solar chimney. Its main structure is the collector – a huge, flat glass roof about 2 meters (6.5 ft.) above the ground, like a vast, low greenhouse. This is open around the sides, and at its center is a tall chimney. The desert sun heats the air and ground, and hot air rises powerfully up the chimney as more air flows in from the sides. The chimney contains an aerogenerator, or wind turbine, to produce electricity. Unlike most other solar generators, the solar chimney could remain active for part of the night, taking advantage of the fact that the ground cools down only slowly after sunset.

A solar chimney 1,000 meters (3,280 ft.) tall, with a glass collector 5 kilometers (3.1 mi.) across, was under construction in 2003 at Mildura, Australia. Once built, such a plant will be almost maintenance-free. It may even be possible to use the outer areas of the collector, where the air is not too hot or fast-flowing, as greenhouses for growing crops. However, as long as petroleum remains relatively cheap and plentiful, solar power is unlikely to take over from oil wells as the major form of wealth provided by the desert. Entrenched interests of governments and corporations and consumer habits make changing energy use an uphill struggle.

## WATER WARS

The absence of water in desert regions may in the future be a potential cause of international conflict, just as the presence of oil and other mineral resources has been a catalyst for war in the past. Mismanagement or overexploitation of this precious resource in one country or region can have catastrophic consequences in a neighboring one, as occurred in the region of the Aral Sea, where extensive irrigation of adjacent semiarid areas has led to the desiccation of vast areas of this formerly great inland sea. In this case, those affected – the fishermen of Kazakhstan and Uzbekistan – are too poor and politically powerless to do very much about their situation. In future, however, such disputes – particularly where they occur between nations – may well escalate into armed conflict.

Access to water is a particularly crucial issue in the Middle East, where settlement of desert areas can serve to reaffirm a country's territorial integrity. Israel's irrigation of the Negev Desert – a triangular parcel of land between Egypt and Jordan – provides not only economic benefits but is a political affirmation of Israel's right to the land. Israel's western neighbor, Egypt, is also concerned with developing settlements in vulnerable marginal areas through large-scale water management, notably in the Sinai Peninsula, which was occupied by Israel from 1967 to 1982. Elsewhere in Egypt the New Valley Project, begun in 1999, aims to divert waters from the Nile to irrigate thousands of square kilometers of desert, both around the fringes of the Nile Delta and into the Sahara's Western Desert. While there are good economic reasons for the project, the "taming" of hitherto uncultivated areas also has political and military benefits, especially in the light of Egypt's at times uneasy relationships with its neighbors – Sudan, Libya, and Israel.

U-shaped mirrors, called heliostats, direct sunlight onto a central collector near Barstow, in the Mojave Desert. The latest project on the site, Solar Two, has proved that solar energy can be an efficient as well as environmentally friendly form of energy generation, and is a prototype for future commercial solar plants.

## A Desert Treasury

Many of the world's deserts early became accessible to large-scale economic exploitation, and the mineral resources of some, like the Great Basin, have already become largely exhausted. At the other end of the scale, a few deserts, such as those of Australia, have to date been only sporadically prospected. Others still, such as the Gobi, appear to be relatively poor in mineral resources.

### American deserts

South America's arid regions hold great metal and mineral wealth. The deserts of western South America have proved especially rich in many valuable metal and mineral resources. Chile has been a world leader in copper production in recent decades – the opencast copper mines around Chuquicamata in the eastern Atacama Desert are some of the world's largest. Many other metals are obtained from the Atacama's fringes, including ores of iron, molybdenum (an element used to harden steel), gold, silver, and lithium (the lightest known metal, used in storage batteries). The Atacama is also a valuable source of nitrates.

In and around the Sechura Desert extraction of copper, silver, zinc, lead, and iron provide more than one-third of Peru's export income. There are also oil fields near Talara on the coast. In Patagonia iron ore and coal are mined, and in 1907 oil was discovered near the seaport of Comodoro Rivadavia.

The Great Basin deserts of North America have yielded many valuable resources, although a large proportion have now been worked out. These include the metals iron, boron, and tungsten, as well as rock salt, soda ash, and sand and gravel for the construction industry. In pre-Columbian times the Aztec people of Mexico mined here for gold, which they used lavishly in the creation of the ceremonial and ritual goods that so impressed the Spanish conquistadores. The Mojave Desert has supplied borax, as well as tungsten, chromium, some silver and gold, and nonmetallic minerals such as salts and potash.

Death Valley was named for its often fatal effects on the fortune hunters who crossed its burning landscape during the Californian Gold Rush of 1848–1849. The valley itself was a valuable source of gold, silver, copper, lead, and in particular borax. Borax was first discovered there in 1873 and brought out by mule trains and their drivers, who labored under extreme conditions made worthwhile by the value of the minerals. Borax is used to make borosilicates for glasses, enamels, lightweight alloys, and, latterly, boron control rods for nuclear reactors. Deserted ghost towns that were once inhabited by the miners are scattered through the region and have become popular tourist attractions.

### Asian deserts

In addition to its immense oil reserves, Saudi Arabia also yields ores of iron, copper, zinc, manganese, tungsten, lead, gold, and silver, as well as phosphates and feldspars. To the north Jordan has resources of salt, potash, phosphates, limestone, gypsum, and marble. Iraq's mineral wealth, apart from petroleum, is of less importance since much of its desert land is high, rocky, and undulating, making transportation and prospecting very difficult.

## DESERT RESOURCES

This table lists the most important resources of each of the major deserts of the world, arranged by continent.

|  |  |
|---|---|
| Atacama | copper, gold, iron, molybdenum, nitrates, silver |
| Sechura | copper, lead, oil, silver, zinc |
| Patagonia | oil |
| Great Basin (inc. Mojave) | boron, iron, salts, tungstun |
| Arabian Desert | chromium, copper, gold, gypsum, iron, lead, oil, manganese, phosphates, salts, silver, tungstun, zinc |
| Kara-Kum | aluminum, copper, iron, lead, natural gas, sulfur, zinc |
| Taklimakan | molybdenum, oil, tungstun |
| Gobi | coal, copper, gold, silver |
| Thar | coal, gypsum, limestone, salts |
| Sahara | iron, mercury, natural gas, oil, phosphates, zinc |
| Namib | uranium |
| Australian deserts | coal, copper, gold, gypsum, iron, lead, nickel, semiprecious stones, zinc |
| Antarctic | oil |

**An oil refinery on the Persian Gulf in Kuwait. Extraction of fuels and minerals has brought increasing industrialization at the margins of the world's deserts.**

Iran is also rich in minerals besides petroleum. Huge reserves of iron ore and coal have led to the development of an important steel industry, and ores of chromium, copper, lead, and zinc are also mined. In September 1980 a long-simmering dispute between Iran and Iraq erupted into war when the latter invaded the former's oil-rich southwestern province of Khūzestān, centered on the port of Abadan, and the desert landscape became the focus of bitter military conflict. An uneasy ceasefire prevailed from 1988.

In the desert regions of Kazakhstan and Turkmenistan are resources of coal, petroleum, natural gas, and ores of iron, chromium, copper, zinc, lead, and aluminum (bauxite). Huge quantities of sulfur are mined at Sernyy Zavod in central Turkmenistan. The Amu Dar'ya, the longest river in Central Asia, flows more than 2,500 kilometers (1,500 mi.) to the Aral Sea and has been dammed for hydroelectricity, with water also drawn off for irrigation along the Kara-Kum Canal.

The Thar Desert of Pakistan and northwest India has been exploited for coal, gypsum, limestone, and salt. The Gobi Desert has fewer known mineral reserves – chiefly some coal, fluorospar, copper, gold, and silver.

### African deserts

As befits its status as the world's largest and most varied desert, the Sahara also furnishes a great range of metals and minerals. Libya and Algeria are both important producers of crude oil, and Algeria also has large natural gas fields, mercury, and zinc, as well as phosphates for fertilizers and the chemical industry. Algeria, Mauritania, and Libya have iron-ore reserves, with ore exports being by far Mauritania's biggest earner. There is also copper in Mauritania, and Morocco is a world leader in the production and export of phosphates.

Uranium ores are mined in Niger's northwest Arlit region, and resources of this metal look likely to be tapped in the Aozou Strip, to the north of the Tibesti Massif. Niger is also developing industries in gypsum, phosphate, salt, and ores of tin, iron, and tungsten, as well as coal and recently discovered petroleum. Mali has largely untapped reserves of iron ore, bauxite (aluminum ore), manganese, gold, marble, limestones, and phosphates.

In Egypt crude petroleum is the leading mineral product, along with salt, phosphates, and ores of iron and manganese. The Nubian Desert of eastern Egypt and Sudan, between the Nile and the Red Sea, furnishes petroleum, ores of iron and manganese, and the hard rocks granite and porphyry for the construction industry. Sudan also possesses ores of iron, copper, manganese, and chromium, as well as gold, gypsum, and some petroleum.

Mineral reserves in the Namib and Kalahari are still being explored. Prospectors here search especially for gold, panning the gravel in streams in the traditional way. There is a large uranium mine at Swakopmund, on Namibia's coast.

### Australian deserts

Australia's desert wealth is still being explored, and looks set for a long and important role in the country's economy. The Great Victoria Desert yields the semiprecious stone opal, especially at the famous "underground town" of Coober Pedy (*see panel*). Coal and iron ore are mined in the Hamersley Range, to the west of the Great Sandy and Gibson deserts. One of the richest iron-ore regions in the world is around the "iron town" of Mount Tom Price. The ore, discovered there in the 1950s, is recovered by opencast methods and taken by train almost 300 kilometers (182 mi.) to Port Dampier, mainly for export.

Other mineral resources occurring in Australia's arid lands include gypsum and the closely related selenite, which are used to make plaster of paris and control the setting speed of portland cement. Nickel is mined in the Kalgoorlie region of Western Australia; ores of lead, copper, and zinc are located north of the Macdonnell Ranges, in an area of the Northern Territory sometimes named the Tanami Desert; and gold is obtained at various sites in Western Australia.

### The Antarctic desert

While the Arctic's lack of large land areas makes exploitation difficult, Antarctica appears to hold enormous mineral wealth. Some 470 million years ago the continent was near the equator and had a rich flora and fauna. It reached its present polar position about 70 million years ago, and is now largely devoid of life (*see* pp. 22–23). Evidence for its former flourishing state remains in the form of fossils and fossil fuels, such as oil and coal.

To date there has been no economic prospecting of Antarctica. In 1991 the Antarctic Treaty nations adopted the Protocol on Environmental Protection, which, in designating the continent a "natural reserve devoted to peace and science," prohibited all mineral activities for at least 50 years, except for the purposes of scientific research. The nations recognize Antarctic's importance, above all, as a wilderness and as an "unpolluted natural laboratory." For the time being, at least, the last great desert is safe.

## UNDERGROUND DOWNUNDER

Some two-thirds of the world's opal is mined around the town of Coober Pedy, some 850 kilometers (525 mi.) north of Adelaide on the eastern edge of the Great Victoria Desert. The Coober Pedy Precious Stones Field of opal and other valuable minerals covers almost 5,000 square kilometers (1,930 sq. mi.) and was first discovered in 1915.

To shelter from the intense heat, early prospectors began the tradition of living in "dugouts" – rooms and chambers excavated into the rock. Life was rough and ready, and many people came to the area to avoid complications elsewhere. Today the town is a tourist attraction, with a population of approximately 4,000 drawn from some 50 nationalities. It has its own water supply and is connected to Australia's infrastructure by the Stuart Highway and daily flights.

Coober Pedy has almost no greenery in the dry season – only low, brown scrub and red desert soil. About 60 percent of the population lives underground, which offers protection from sandstorms, in dwellings dug horizontally into the sandstone hillsides. There are some 800 houses (*right*, for example) as well as shops, motels, and churches, below ground. Inside these "dugouts" the temperature is a steady 20–25°C (68–77°F) all year round. In the summer surface temperatures regularly exceed 40°C (104°F) and may break 45°C (113°F), while winter nights can be close to freezing. The name of Coober Pedy is said to be a corruption of the Aboriginal name for the settlement – *Kuba Piti*, meaning "white man's burrow."

# SPREADING DESERTS

*Around the world deserts are spreading. The Atacama Desert in Chile grows by an area of more than 10 soccer fields each day; the Sahara is expanding southward at 10 times this rate. In many cases the new deserts are not true desert, but poor-quality drylands.*

Among the environmental concerns that have occupied scientists, the public, and the media in recent years, what is loosely – though somewhat misleadingly – termed the problem of spreading deserts and more correctly called desertification has often taken a back seat. Unlike such headline-grabbing issues as global warming, nuclear waste, and pollution, which are generally considered to be of global importance, desertification affects mostly the poor and poorer countries. As with these other issues, though, scientists hotly debate the extent and relative seriousness of desertification.

The United Nations (UN) defines desertification as the "degradation of dryland [that is, semiarid] areas" and considers it of sufficient global importance to establish a Convention to Combat Desertification, of which the first session met in Rome in 1997 and which to date has 110 national signatories. According to UN statistics, some 70 percent of the earth's drylands – excluding hyperarid desert areas – are degraded, affecting 250 million people worldwide, with a further billion people at risk of becoming affected. The primary victims of desertification suffer food and water shortages, poverty, and, in the worst cases, famine and mass migration. People living in adjoining areas to those affected by desertification also suffer problems, from increased health problems to conflicts arising from migration and the struggle for scarce resources.

Desertification is considered a major global environmental issue largely because of the close link between dryland degradation and food production. Maintaining a nutritionally adequate diet for the world's rapidly growing population entails a tripling of food production over the next 50 years. This will be difficult to achieve even under favorable circumstances, and if the process of desertification is not stopped and reversed, food yields in many affected areas will decline. Global malnutrition, starvation, and ultimately famine may result.

**The gray, desiccated remains of trees show the effects of desertification in Western Australia, near Perth. Wholesale land clearance and intensive crop cultivation have led to rapid soil erosion in many semiarid parts of the continent; some experts estimate that for every kilogram (2.2 lb.) of bread produced in Australia, 7 kilograms (15.4 lb.) of top soil are lost forever.**

## Natural and Unnatural Deserts

The expansion of desert areas is not always the product of human intervention. Geographers draw a distinction between desertification – the degradation of semiarid areas, sometimes in regions not usually thought of as desert zones, due to, for example, overcropping or poor water management – and the natural process of desertization – the growth, shrinkage, and movement of deserts that has taken place over millions of years in response to global climate change (*see* Chapter 1).

### Desertization

Geological studies of rocks and fossils reveal how desert regions have waxed and waned. They reveal, for example, that the Sahara has become progressively drier since the end of the last ice age some 10,000 years ago. Before that time, when much of northern Europe was covered by ice sheets, the northern stretches of Africa were cooler and damper than they are today, crisscrossed by rivers carrying moisture delivered by the North Atlantic weather systems. These ancient rivers are often visible in satellite photography.

As the ice age faded, however, these rain-carrying air currents shifted northward, roughly to the latitudes of Britain and France. Even 6,000 years ago grasses and low shrubs across North Africa provided a equable habitat for both animals and people. As the hot, dry winds increased, the great, fiercesome desert we know today began to take a grip. Computerized climate modeling suggests that a critical point was reached some 5,500 years ago when, deprived of its binding moisture, the Saharan surface broke down within perhaps as little as half a millennium. By 4,000 years ago the sterile sea of sand and rock had become established.

Natural changes in the extent of deserts are driven by global climate fluctuations. They usually occur slowly, giving plants and animals time to adapt, or to move on to more hospitable habitats.

### Desertification

Despite temporary fluctuations in its fragile mix of surface features, plants, and animal life, the natural desert is a relatively stable and balanced environment. Drylands, or semiarid lands, which are often situated on the fringes of hyperarid, true deserts, are altogether more fragile, subject as they are to uncertain rainfall. Over the ages, dryland ecology has become attuned to this variability in moisture, with plants and animals responding rapidly to change. Satellite imagery has shown, for example, that the vegetation boundary south of the Sahara can shift by up to 200 kilometers (124 mi.) when a wet year is followed by a dry one, or vice versa. Traditionally humans, too, have learned to protect the drylands, adopting strategies such as shifting agriculture and nomadic herding and in general responding flexibly to climate fluctuations.

In recent decades, however, the pressures of population growth and economic and political change have made it increasingly hard for local people to sustain such strategies. Settlements have grown, leading to unsustainable land use that fails to respond flexibly to environmental fluctuations, such as overgrazing and poor irrigation practices. The result – desertification – is altogether swifter and more devastating than the process of desertization, leading to the creation, not of a true desert, but of a deserted, desolate landscape (*see* Fig. 1).

The broader causes of desertification are frequently rooted in local, national, and international inequalities that lead to poor land management. In many instances rich individuals

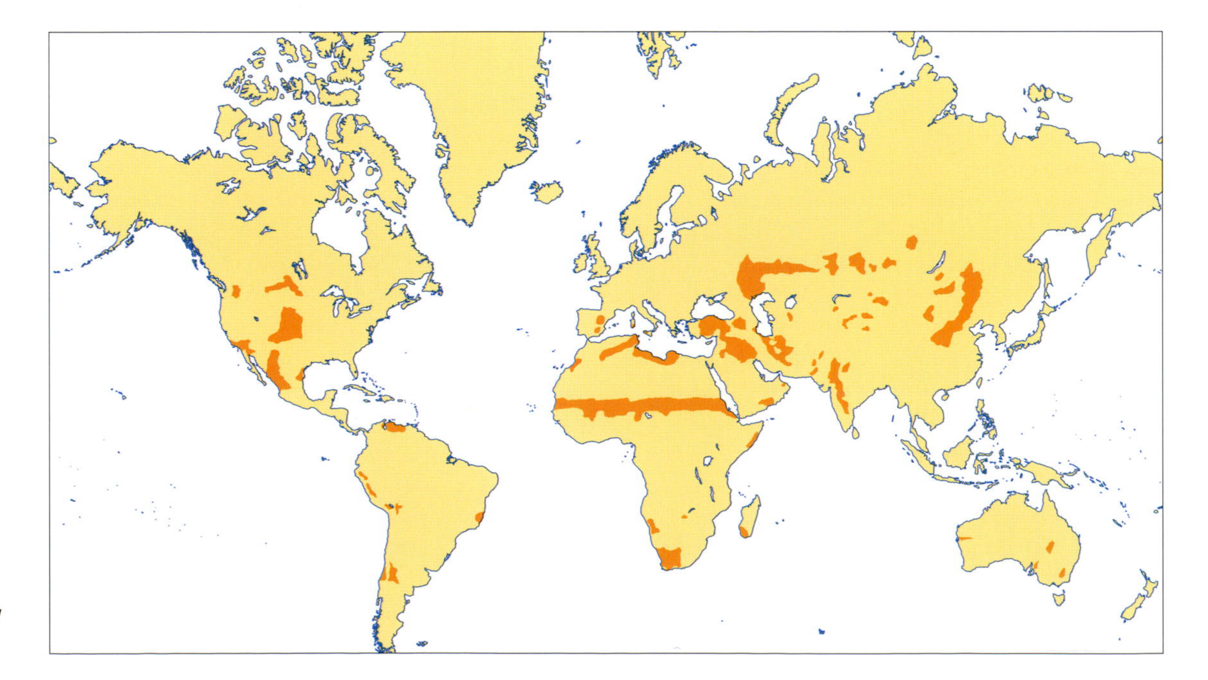

**Figure 1 Areas of the world undergoing desertification, measured by degradation of soil quality. Most such areas are in close proximity to the world's deserts.**

In the dryland Saharan fringes of Mauritania desertification has led to the abandonment of villages, such as the one shown here. The Sahara expands and shrinks naturally, but in many instances, as here, poor land management leads to irreversible degradation of lands that previously supported nomadic pastoral and agricultural lifestyles.

within a community maximize their own profits by overexploiting the land at the expense of the community as a whole. Poorer people often lack access to the best land and depend instead on the most fragile areas. Their poverty may force them to extract what they can from the scarce resources available to them, even at the cost of the long-term sustainability of the land.

International economic forces can also encourage people to overexploit their land. International trade patterns that favor richer, industrialized nations at the expense of poorer, less developed ones can lead to the unsustainable exploitation of local resources for export, as well as leaving little profit at the community level for managing or preserving the land. Similarly, the development of an economy based on cash crops, or the imposition of taxes, can distort local markets and promote overexploitation of the land.

## GOOD MANAGEMENT

The collection of wood for fuel is a major factor in desertification on desert fringes, where expanding populations put increasing pressure on the low local biomass. However, the connection between population growth and environmental degradation is not inevitable, particularly where local people are able to take control of local land use. In Kenya a six-year-long survey in the 1990s showed that the human population increased by some 24 percent, yet the woody volume of planted trees grew by almost 30 percent. Kenyans, it seemed, had planted more trees than they had used, often as part of well-managed, sustainable projects, in which farmers used a traditional mixture of food crops and cash crops, livestock grazing, and tree-planting. The Kenyan Green Belt Movement, funded in part by the United Kingdom's Comic Relief charity, has planted more than 20 million trees, encouraging especially women in rural farming areas to take part in the project through the payment of small cash amounts for each tree established.

The problem of desertification is occurring in more than a hundred countries around the world; some experts suggest that as much as one-third of all ice-free land area is at risk. Global annual costs of desertification are put at $50 billion, but the primary human cost is borne by millions of already poor and marginalized people, for whom the withering of the land is a catastrophic, even life-threatening, event, and who have little political or financial clout to do anything about it.

Even the most conservative estimates suggest that about 7 million hectares (17 million acres) of land around the world have changed from "dry rangeland," where some grazing or agriculture is possible, to almost worthless desertlike conditions. Another 5 million hectares (12 million acres) of better-watered farmlands have gone the same way, overcropped and unsuitably managed – many of them in rich countries such as the United States. Another recent survey indicates that of the planet's total ice-free land area, some 6 percent is already hyperarid true desert, while almost the same area is at high risk of desertification.

### Causes and Consequences

Plants play a vital role in the maintenance of any ecosystem – especially so in fragile, unstable environments such as drylands – and the disturbance or loss of plant cover, leading to reduced soil fertility, is the major triggering factor in desertification. Three major human activities are crucial here – livestock grazing, crop cultivation, and wood collection – and irresponsible or mismanaged undertaking of these can all too easily destroy vulnerable plant life.

The degree of damage caused to habitats from grazing livestock is dependent on type of livestock as well as on the quality of grazing management. Cattle and sheep are relatively choosy grazers and tend to nibble off grassy blades, leaving the stems and roots to regenerate. Pigs are more omnivorous, disturbing the soil structure as they grub for roots and tubers and also taking valuable soil-enriching creatures such as worms. Goats are an even greater threat to plants. Famed for their ability to eat almost anything – including tough bark, thorny stems, and prickly leaves – as well as for their climbing ability, they can feed among rocks and boulders and even leap into low branches to bite off shoots and buds.

Crop cultivation is another threat to soil fertility around the world. Dryland soils are usually sparse and poor in minerals and nutrients, but careful planting, enrichment, and harvesting may coax the occasional yield of dry-adapted crops. More typical today, however, are farming methods in which short-term gain triumphs over long-term viability. Wholesale plowing and similar mechanical treatments, together with the soaking of thin soil in artificial fertilizers and chemical pesticides, may well produce more profitable harvests for a few years, but will eventually upset the natural cycle by which the soil renews its fertility.

Wood is used mainly as a biofuel – burned as firewood for heat, especially for cooking and boiling water, and for light – but trees are also harvested for timber to enable the construction of buildings, bridges, and other structures. Even the occasional branch chopped off a living tree weakens its resistance to drought, and the open wound is an invitation to disease.

**Such startling sights as a golf course in the midst of desert dunes may seem miraculous but are usually created at the cost of long-term environmental damage. This lush green expanse, in the United Arab Emirates, depends on copious water drawn from the water table far below the desert surface, a precious resource that only slowly replenishes itself.**

## Degradation and erosion

As plants wither and die above ground, their roots and other underground parts shrivel. The land loses a major stabilizing influence below ground as well as its chief protective layer above the surface. Without stems and leaves working as sunshades and windbreaks, the heat and breeze begin to dry out the soil's surface. Even when it does rain, the water droplets pound straight onto the bare earth, rapidly turning it into a mobile slurry. In the absence of plants to function as living sponges and suck up moisture, excess water collects and pools soil particles into cakelike sediments. Without roots to act as mudbreaks and slow the flow of subsurface water, runoff is accelerated.

The next phase is the physical removal of the soil from the land. This can happen with alarming speed during a heavy rainstorm when within minutes the degraded, rootless soil loses its texture and integrity and begins to slide toward lower ground, into channels and runoffs. As the water accumulates, its pressure, flow, and soil-carrying capacity increase. After an hour or two the top layer of soil is on the move, flowing away to clog up rivers, lakes, and waterways downstream.

This process of soil degradation and loss can happen anywhere. It is especially spectacular in rain-forest areas that have been logged or cleared for farmland, where one torrential downpour can turn a hill formerly cloaked in dense vegetation and rich leaf litter into a pile of mud and fallen,

## THE ARAL SEA

Intensive irrigation can dramatically increase the agricultural productivity of dryland areas, but often at high environmental cost. One of the greatest ecological catastrophes of modern times is the desertification of the Aral Sea in Central Asia, consequent to massive Soviet irrigation projects begun in the 1960s, which diverted water from the sea's tributary rivers to maintain cotton fields established in dryland areas of Kazakhstan. Previously the Aral Sea had been the earth's fourth-largest body of inland water and a rich resource for local populations. Since the 1960s the sea has shrunk to nearly half its former size. As the shoreline has retreated, once-prosperous fishing ports have been left stranded, sometimes as far as 60 kilometers (40 mi.) from the sea, and large desertlike areas have formed, whose pesticide-laden dust is a serious health hazard for impoverished local communities. Improving the quality of the irrigation channels may stop much of the water loss, but the degradation of large areas of the Aral Sea seems irreversible.

matchsticklike trees. In drylands soil erosion is usually less obvious because the soil itself is both rarer and thinner, but it is just as devastating.

### Unhelpful introductions

The introduction of dry-adapted animals and plants from their native lands can also cause problems for dryland habitats and contribute to desertification. One cautionary tale that is particularly well documented is that of the introduction of and devastation caused by the humble rabbit to the arid Australian outback.

Early European settlers took this prolific, rapacious mammal to Australia, partly as a familiar food supply and partly as an attempt to domesticate what seemed an alien landscape, to make the outback "more like home." In the late 1850s a Mediterranean variety of rabbit began to spread unchecked from Victoria and within 30 years had migrated across the arid inland areas of all the mainland states.

Thousands of kilometers of rabbitproof fences were erected to no avail: vast tracts of dry pastureland, already overgrazed by sheep first introduced to the continent in 1797, were grazed, nibbled, burrowed, and eroded into dust. Spin-off destruction included the decimation of native Australian species such as bettongs and wombats, which were heavily outcompeted. The furry pest reached plague proportions, with one farmer reported to have killed 30,000 rabbits in four days around his ranch and yards. Not until the viral disease of myxomatosis was deliberately introduced for a second time in 1950 was Australia's rabbit "plague" brought under control.

### Mismanaging water

The mismanagement and overexploitation of water is another major cause of desertification. Grandiose but short-sighted irrigation schemes in semiarid areas are rarely sustainable and may cause environmental catastrophe (*see panel above*).

Water mismanagement extends to the desert itself, where few sights are more striking than an eruption of greenery among barren dunes. In the case of palm-fringed oases this can be a natural occurrence, although water production is sometimes speeded by mechanical pumping; altogether

## THE SAHEL

To its southwest the Sahara shades into a vast dryland area of bush, scrub, and shrub known as the Sahel, extending from southern Mauritania and Senegal east through Burkina Faso and Niger to northeast Nigeria and neighboring Chad and west Sudan. To the southwest of the Sahel is the tropical forest belt of West Africa, and to its southeast the scrub and savannas of East Africa.

The Sahel's annual rainfall varies from 100 to 600 millimeters (4–23.5 in.), falling mainly from June to August, with a prolonged eight-month dry spell. Traditionally Tuareg, Fulani, and other nomadic peoples have occupied the region's northern reaches, while farther south the thin soil provides grasses for grazing camels, cattle, and sheep, as well as nutrients for raising a few crops such as millet and groundnuts.

In the Sahel desertization and desertification have combined. Climate records suggest a long-term decline in rainfall, with severe droughts recorded in the mid-1910s, early 1940s, from 1968 to 1974, and sporadically throughout the 1980s. This, along with the prevailing sand-blowing winds, has encouraged the Sahara to spread southward. However, overgrazing and unsuitable agriculture have also played their part, as the people of the region struggle simply to survive. Fuelwood has become scarce, leading to the chopping down of the few remaining trees.

Recent human history has also played its part. During the 1950s and early 1960s the Sahel had better-than-average rainfall. Encouraged by the short-term success of cash-crop production in regions farther south, local people abandoned their traditional lifestyle mix of nomadic pastoralism and shifting agriculture for a more intensive cultivation of the land. When drought struck in 1968, the effects were all the more harsh as a result of the changeover, with an estimated 100,000 people dying within three years.

There is some hope, however. Studies undertaken in the Sahel during the 1990s have shown that in some areas the population has returned to the traditional mix of herding and subsistence agriculture. Many people now have sufficient crops in most years, and some are trying out their own hybrids of domesticated sorghum and millet and tougher, more disease-resistant wild varieties.

unnatural, however, are emerald-green, sprinkler-studded golf courses and well-watered stands of fruit trees. Both phenomena – natural and unnatural – owe their existence to the vast reservoirs of water that lie under many deserts, confined within layers of rain-soaked rocks called aquifers. Tapped by humankind for thousands of years, in the form of natural springs and oases as well as artificial wells and boreholes, these natural reservoirs are now subject to a more wholesale exploitation, threatening their long-term viability.

Groundwater takes thousands of years to accumulate in its catchment area, perhaps hundreds of kilometers away from where the rains fall, and intensive abstraction uses it up hundreds of times faster than it can be naturally replenished. Sometimes, of course, groundwater is used for less profligate reasons than the laying-out of golf courses and orchards; for example, the planting of native dry-adapted species is used as a means of greening the arid surface and halting any spread of the desert. However, all such groundwater-

**Drifts of dry, windblown soil pile up against an abandoned homestead. During the 1930s, whole areas of central North America suffered catastrophic desertification. Fragile farmlands that had been arduously plowed and planted by generations of pioneering farmers became barren wasteland in a matter of months.**

abstraction projects are unsustainable, even in the medium term. As water is sucked out and the water table falls, the wells and boreholes run dry, or salts and other minerals concentrate to such a degree as to make both soil and water unusable. In addition, abstraction of groundwater can cause subsoil and bedrock to crack and subside, with consequent damage to the surface and to structures such as roads, bridges, railroads, canals, and pipelines.

### Another cautionary tale

Several once-fertile farming regions have been reduced to barren wastelands by combinations of overcropping and overgrazing. Most infamous was the 1930s Dust Bowl in North America, stretching from southern Colorado and Kansas through eastern New Mexico across Oklahoma to Texas. Some 400,000 square kilometers (155,000 sq. mi.) of former prairieland, which in the 19th and early 20th century had been converted into farmland, was turned into dusty wasteland.

In part, this environmental catastrophe was a natural occurrence. The region's relative aridity – with annual rainfall as little as 400 millimeters (15 in.) in places – was compounded by failed rains in the years from 1934 to 1937, leading to severe drought. However, the main culprit was the poor farming methods that had already reduced the soil's fertility. The use of early motorized farm vehicles such as tractors and combine harvesters, combined with government assistance to farmers, enabled the rapid and ill-considered cultivation of this vast region.

Lacking protection from the strong local winds and the anchorage of grassy roots, the naturally light soils dried and disintegrated into fine dust. This was picked up by the winds as swirling "black blizzards" that blanketed the landscape, with vehicles and even whole houses disappearing beneath huge piles of falling dust. More than half the population left the area, their troubles later immortalized in John Steinbeck's novel *The Grapes of Wrath*, published in 1939.

This near-legendary American disaster serves as an illustration of what may lie in store for many of the world's degraded arid lands. Parts of South America and northern China suffered smaller-scale "dust bowls" in the 1970s, and similar episodes are doubtless brewing elsewhere. Unlike the United States, which was wealthy enough to repair much of the damage of its own Dust Bowl, those countries most likely to suffer this environmental catastrophe today will probably be too poor even to begin reversing its effects.

## The Future

Desertification is not inevitable and can be slowed and even reversed. Several projects have recorded major success in Australia, Mali, Nigeria, and Chile. Most have involved planting programs that focus on native species, which, naturally drought-adapted, have begun to stabilize the soil, raised its moisture and humus content, and gradually encouraged other species to return. The erection of wind–sand barriers of feather, sage, and similar grasses, and hardy trees such as acacias, has also achieved some encouraging results. These are minor triumphs, however, in a massive – and high-risk – global war.

**In Mauritania the Fight the Desert project uses simple strategies to prevent the spread of the desert. Fences of wood from a local tree are placed in 30-by-30 meter (98 x 98 foot) squares to prevent the sand from blowing away, thereby stabilizing the desert soil.**

### Climate change

During the 1970s and 1980s several major discoveries were made concerning global atmospheric pollution, including acid rain, ozone depletion, and the "greenhouse effect" of heat-retaining gases in the atmosphere leading to global warming. Many scientists have predicted climate change, with early trends detected from the 1990s not only of raised temperatures but also an increased frequency of violent atmospheric events such as hurricanes, typhoons, storms, and floods.

If worst-case predictions come to pass, climate change will almost certainly affect the distribution and extent of deserts. One climatic model suggests that the dry belts of high-atmospheric pressure at subtropical latitudes about 25–30° north and south of the equator, which now maintain most of the world's deserts, may shift to slightly higher latitudes. This may bring drier conditions nearer to densely populated regions, such as those situated around cities such as San Francisco, Chicago, New York, Rome, Beijing, and, in the southern hemisphere, Montevideo, Cape Town, and Sydney. Whether these areas will irresistibly become "deserts," while today's deserts turn leafy and green, it is too early to say.

# Glossary

**allogenic** *See*: **exotic**.
**alluvial fan** A fan-shaped alluvial deposit of a stream or river where it emerges from a gorge.
**alluvium** Silt, gravel, and similar detrital material deposited by running water.
**aquifer** A permeable rock layer that absorbs water.
**aridity** Quality defining climate or landscape that lacks sufficient precipitation to support general vegetation.
**arroyo** A Spanish term applied to watercourses in arid or semiarid areas in North America.
**badland(s)** Semiarid terrain marked by erosion and scant vegetation.
**barchan** A moving crescent-shaped sand dune formed where a wind blows constantly from one direction.
**bedrock** The solid, relatively unweathered rock that lies beneath a surface such as soil and eroded rock.
**Benguela current** A cold current that flows northward along the western coast of southern Africa and helps maintain the cold-water Namib Desert.
**biomass** The amount of living matter that can be supported by a particular habitat.
**biome** A major habitat type (e.g., rain forest, desert, and tundra).
**butte** A flat-topped, steep-sided rocky outcrop, smaller than a mesa, found in arid and semiarid regions.
**capillary action (capillarity)** The mechanism by which water is attracted up through rock and soil, often carrying deposits such as salt to the surface.
**chemical weathering** The breakdown of rocks through chemical processes such as oxidation and hydrolysis.
**chott (shott)** Arabic term for a shallow salt lake that dries out in summer and which is often fed by an underground water source.
**cold desert** Term applied to arid regions at higher altitudes to the north and south of subtropical, "hot" deserts, often on the leeward side of mountain ranges or deep within a continent.
**cold-water desert** Term used to a describe a coastal desert whose western edges are washed by a cold ocean current, producing lower mean annual temperatures than normally found at such latitudes.
**Coriolis effect,** or **force** The deflection caused by the rotation of the earth; an important phenomenon in the creation of the world's air and water currents, and hence of some kinds of deserts.
**creek** A seasonal stream in the arid American Southwest and inland Australia. *See also*: **arroyo**.
**cryophyte** A plant adapted to living in permanent snow or ice.
***dasht*** or ***desht*** Persian word for "desert" or "plain."
**degradation** In soil science a term used to describe excessive weathering and leaching of nutrients, often due to poor agricultural practices.
**desert** An arid region with little or no rainfall and sparse vegetation. *See also*: **cold desert**; **cold-water desert**; **hot desert**; **polar desert**; **semidesert**.
***désert*** French word for "desert."
**desert pavement** Large area of pebbles and larger stones in an arid environment that have been polished by the wind to form a smooth, pavementlike expanse. *See also*: **hamada**.
**desert varnish** A hard shiny glaze that forms on exposed rock surfaces in deserts, the result of iron or magnesium oxides being drawn to the surface by capillary action.
**desertification** The degradation of typically semiarid areas, or drylands, due to human interference such as overgrazing and poor irrigation practices.
**desertization** The waxing and waning of desert areas due to natural causes such as climate change.
**desiccation** An increase in aridity.
***desierto*** Spanish word for "desert."
**dominant wind** The wind that plays the greatest role in shaping a local environment. *Compare*: **prevailing wind**.
**drâa** Arabic term applied to the largest accumulation of sand in the Sahara Desert; "sand mountain."
**drought** A prolonged period of zero or very low rainfall, usually defined differently in different countries.
**drylands** Semiarid lands.
**dune** In hot deserts a mound or other formation of windblown sand. *See also*: **barchan**; **kopi dune**; **seif dune**; **star dune**; **transverse dune**.
**dust bowl** An area of semiarid land that has lost most of its soil owing to overcultivation or overgrazing.
**dust devil** A short-lived whirl of dust in desert areas caused by intense local solar heating of the surface creating a powerful convection current.
**dust storm** A hot desert or dryland storm that whips up large quantities of sand.
**eolian** Term used to describe landscape forms shaped by the wind.
**ephemeral stream** A watercourse in an arid or semiarid environment that flows only after heavy rainfall.
**erg** Arabic term applied to a large area of sand dunes; "sand sea."
**evaporation** The transformation of liquid into gas.
**exfoliation** The splitting and peeling-off of outer layers of rock.
**exotic** Term used to describe a river that arises from a water source in a different climate zone and can flow through desert without drying up. *Also called:* **allogenic**.
**flash flood** A sudden flood due to heavy rainfall.
***gobi*** Mongolian word meaning "arid steppe" or "desert."
**groundwater** Water in cracks and pores of rocks, deriving from atmospheric precipitation or from deep magmatic sources, and which supplies springs, wells, and oases.
**Hadley Cell** Tropical part of the global atmospheric

circulation pattern, including the trade winds; named after the 18th-century English scientist George Hadley.
**halophyte** Term used to describe plants that tolerate salt.
**hamada** Arabic term used to describe bare-rock desert.
**harmattan** A dry wind that blows southwest from the Sahara across the Sahel.
**hoodoo** A type of oddly shaped rock pinnacle in semiarid North America, resulting from undercutting by wind.
**hot desert** Term sometimes applied to subtropical, high-pressure climate deserts.
**inselberg** A steep-sided rocky outcrop on a low-lying plain, often found in desert landscapes.
**insolation** The relative amount of solar radiation that reaches the earth's surface; equatorial insolation is some 2.5 times greater than that at the poles.
**khamsin** Arabic term for the hot, dry wind that sometimes blows across Egypt from the Sahara.
**kopi dune** Linear dunes in semiarid Australian inland featuring gypseous deposits.
***kum*** *See*: ***qum***.
**longitudinal dune** *See*: **seif dune**.
**mechanical weathering** The breakdown of rocks by weathering agents such as wind and running water. *Also called*: **physical weathering**.
**mesa** Spanish term applied to an isolated flat-topped hill or rocky outcrop. *Compare*: **butte** *and* **tableland**.
**mulga** Scrubby vegetation type found in arid Australia.
**nullah** A dry watercourse flowing only during the monsoon rains in arid and semiarid India.
**oasis** A fertile area in a desert watered by a natural spring or artesian well or borehole.
***oued*** Arabic word for "valley" or "dry watercourse"; wadi.
**oxidation** A chemical weathering in which oxygen dissolved in water reacts with a mineral such as iron, creating a brownish or yellowish stain on the surface of the rock.
**paleontology** The scientific study of fossils.
**pan** A land depression that sometimes fills with dried-up deposits such as salt or clay.
**Pangaea ("All-Earth")** Name given to the hypothetical primeval landmass comprising the present continents.
**pedestal** A small mushroom-shaped rock formation created by wind erosion in arid and semiarid regions.
**pediment** A gentle slope stretching from a steeper, often mountainous, slope down to a river or alluvial plain.
**Peru current** A cold current that flows northward along the western coast of South America and helps maintain the cold-water Atacama and Sechura deserts.
**physical weathering** *See*: **mechanical weathering**.
**plateau** A tract of elevated land.
**playa** Spanish term applied to a level area at the center of a basin in which a lake periodically forms; often characterized by stratified layers of dried deposits such as salt or clay.
**precipitation** The deposition of solid or liquid water on the earth from the atmosphere (e.g. snow and rain).
**polar desert** High-latitude areas where cold and dryness severely inhibit plant growth.
**prevailing wind** The wind that blows most often in a certain region. *Compare*: **dominant wind**.
**psammophyte** A plant adapted to live in shifting sand.
***pustynya*** Russian word for "desert."
**qaid** A range of roughly conical sand dunes.
***qum*** Turkic word for "desert." Also *kum*.
**rain-shadow** The area on the leeward side of mountains that has low rainfall because rain-bearing winds have already given out their moisture on the windward side.
***ramlat*** Arabic word meaning "sandy area."
***ran*** Hindi word for "wilderness."
**reg** Arabic term applied to flat, stony desert, in which most of the finer particles have been removed by the wind.
***sahra (pl. sahara)*** Arabic word for "desert."
**salar** Term applied to a playa in the American Southwest.
**salt lake** A lake that forms in desert depressions, with high salinity owing to rapid water evaporation.
**salt pan** A small, shallow salt lake.
**saltation** The bouncing movement of windblown particles across a landscape; the most common form of wind transport in arid regions.
**sebkha** Arabic term applied to an arid, salt-filled depression, often in coastal regions close to a desert, as in the far southeastern corner of the Arabian Peninsula.
**seif dune ("sword dune")** A sharply crested sand ridge.
**semiarid** Term loosely used to describe regions with only light rainfall, sometimes defined specifically as 25 to 50 centimeters (10–20 in.) of annual precipitation.
**semidesert** Term applied to arid and semiarid transitional regions between desert and more thickly vegetated environments.
**shott** *See*: **chott**.
**sirocco** A hot, dusty wind that blows from the Sahara into Europe, gathering moisture.
**soil erosion** The removal of soil at a greater rate than it can be replaced by natural agencies.
**star dune** Massive pyramid-shaped sand dune with radiating sand ridges.
**subtropics** Areas immediately north and south of the tropics of Cancer and Capricorn respectively, often characterized by a warm, dry climate.
**tableland** Flat-topped or undulating land of high relief.
**takyr** Russian term applied to a wind-polished clay-floored depression in which a seasonal lake forms.
**transverse dune** Irregular dune that forms at right angles to the direction of the wind.
**wadi** A river or stream-formed valley in arid or semiarid areas of North Africa and the Arabian Peninsula; wadis are seasonally or permanently dry, but after heavy rainfall may carry torrential flows of water. *Compare*: **arroyo**.
**xerophyte** Term applied to plants that are dry-adapted.
**yardang** A sharp-crested linear rock ridge found in a desert environment.

# Bibliography


**General sources**

Bagnold, R.H. *The Physics of Blown Sand and Desert Dunes.* London, UK: Methuen (1941).

Botkin, D., and E. Keller. *Environmental Science: Earth as a Living Planet.* New York, NY: John Wiley (1995).

Chapman, V.J. *Salt Marshes and Salt Deserts of the World.* New York, NY: Interscience Publishers (1960).

Holmes, A. *Principles of Physical Geography.* Cheltenham, UK: Thomas Nelson & Sons (1944).

Lake, Philip. *Physical Geography.* Cambridge, UK: Cambridge University Press (1965).

Mares, Michel A. (ed.). *Encyclopedia of Deserts.* Norman, OK: University of Oklahoma Press (1999).

*New Encyclopaedia Britannica,* 15th edn. Chicago, IL: Encyclopaedia Britannica (1995).

Petrov, M.P. *Deserts of the World.* New York, NY: John Wiley (1976).

Pye, Kenneth, and Haim Tsoar. *Aeolian Sand and Sand Dunes.* Dordrecht, Netherlands: Kluwer Academic Publishers (1990).

*Quaternary Deserts and Climatic Change: Proceedings of an International Conference, Al Ain, 9–11 December 1995.* Rotterdam, Netherlands: A.A. Balkema Publishers (1998).

Skujins, J. (ed.). *Semiarid Lands and Deserts.* New York, NY: Marcel Dekker (1991).

Wilde, Peter (ed.). *Desert Reader.* Salt Lake City, UT: University of Utah Press (1991).

*The Sand Seas of Earth*
http://syninfo.com/Crystal/xls012.htm

*The World's Biomes: Deserts*
http://www.ucmp.berkeley.edu/glossary/gloss5/biome/deserts.html

**Chapter 1: How Deserts Form**
**Atlas: African Deserts**

Brookfield, M.E., and T.S. Ahlbrandt (eds). *Eolian Sediments and Processes.* Amsterdam, Netherlands: Elsevier (1983).

Dennis, Nigel, Michael Knight, and Peter Joyce. *The Kalahari: Survival in a Thirstland Wilderness.* Cape Town: Struik Publishers (1997).

Durou, Jean Marc. *Sahara: The Forbidding Sand.* New York, NY: Harry N. Abrams (2000).

Fleming, Fergus. *The Sword and the Cross: The Conquest of the Sahara.* London, UK: Granta Books (2003).

Gerster, G. *Sahara: Desert of Destiny.* New York, NY: Coward-McCann (1960).

Glennie, K. W. *Desert Sedimentary Environments.* New York, NY: Elsevier Publishing (1970).

Kennedy, Geraldine. *Harmattan: A Journey across the Sahara.* Santa Monica, CA: Clover Park Press (1994).

Lancaster, N. *The Namib Sand Sea: Dune Forms, Processes, and Sediments.* Rotterdam, Netherlands: A.A. Balkema Publishers (1989).

Langewiesche, William. *Sahara Unveiled: A Journey across the Desert.* Reprint edn. New York, NY: Vintage Books (1997).

Lanting, Frans. *Okavango: Africa's Last Eden.* San Francisco, CA: Chronicle Books (1995).

Luard, Nicholas. *The Last Wilderness: A Journey across the Great Kalahari Desert.* New York, NY: Simon & Schuster (1981).

Martin, Michael. *Deserts of Africa.* New York, NY: Stewart, Tabori, & Chang (2000).

Owens, Mark, and Delia Owens. *Cry of the Kalahari.* Reissue edn. Boston, MA: Houghton Mifflin (1992).

Swift, J. *The Sahara.* Amsterdam, Netherlands: Time-Life International (1975).

Van der Post, Laurens. *The Lost World of the Kalahari.* Harmondsworth, UK: Penguin (1977).

*Deserts and Desert Processes*
http://www.ux1.eiu.edu/~cfjps/1300/deserts.html

*Deserts and Wind*
http://www.clas.ufl.edu/users/dfoster/courses/2010_25_notes.pdf

*Dunes Geology*
http://www.glamisonline.org/dunes.asp

*Eolian Processes*
http://pubs.usgs.gov/gip/deserts/eolian/

Nelson, Stephen A. *Wind Action and Deserts*
http://www.tulane.edu/~sanelson/geol111/deserts.htm

**Chapter 2: Sand, Rock, and Rubble**
**Atlas: Asian Deserts**

Blackmore, Charles. *Crossing the Desert of Death: Through the Fiercesome Taklamakan.* New edn. London, UK: John Murray (2000).

Facklan, Margery. *Tracking Dinosaurs in the Gobi.* New York, NY: Twenty-First Century Books (1997).

Glennie, K.W. *Desert Sedimentary Environments.* New York, NY: Elsevier Publishing (1970).

Hedin, Sven. *Across the Gobi Desert.* Oxford, UK: Greenwood Press (1968).

Herrmann, Georgina. *Monuments of Merv: Traditional Buildings of the Karakum.* London, UK: Society of Antiquaries of London (1999).

Philby, H. St. John B. *The Empty Quarter: Being a Description of the Great South Desert of Arabia Known as Rub' Al Khali.* New York, NY: Henry Holt (1933).

Sharma, R.C. *Thar: The Great Indian Desert South.* New Delhi: Roli Books (1998).

Xuncheng, Xia. *Wondrous Taklimakan.* Tokyo: Science Press (1993).

*Desert Features*
http://pubs.usgs.gov/gip/deserts/features/

**Chapter 3: Plants of the Desert**

Batanouny, K.H. *Plants in the Deserts of the Middle East: Adaptions of Desert Organisms.* Hamburg and Berlin, Germany: Springer-Verlag (2001).

Dodge, Natt Noyes. *Flowers of the Southwest Deserts*. Tucson, AZ: Southwest Parks & Monuments Association (1985).

Houk, Rose. *Mojave Deserts: American Deserts Handbook*. Tucson, AZ: Southwest Parks & Monuments Association (2001).

Murray, John. *Cactus Country: The Deserts of the American Southwest*. Boulder, CO: Robert Rinehart Publishers (1996).

Pickering Larson, Peggy, and others. *The Deserts of the Southwest: A Sierra Club Naturalist's Guide*. 2nd edn. San Francisco, CA: Sierra Club (2000).

Tivy, Joy. *Biogeography*. 3rd edn. London, UK: Longman Group (1995).

American Desert Plants: *The Living Desert* http://www.livingdesert.org/

**Chapter 4: Creatures of the Desert**
**Atlas: American Deserts**

Bender, Gordon L. *Reference Handbook on the Deserts of North America*. Westport, CT: Greenwood Publishing Group (1982).

Chatwin, Bruce. *In Patagonia*. New York, NY: Vintage Books (1998).

Cornett, James W. *Atacama: Desert of Chile and Peru*. Palm Springs, CA: Palm Springs Desert Museum (1985).

Neilson, David (photographer). *Patagonia: Images of a Wild Land*. Emerald, Vic.: Snowgum Press (1999).

Savage, Stephen. *Animals of the Desert*. Austin, TX: Raintree Steck-Vaughn (1997).

Wuerthener, George (photographer). *California's Wilderness Areas: The Complete Guide. Vol. 2: The Deserts*. Englewood, CO: Westcliffe Publishers (1998).

*Desert USA* http://www.desertusa.com/survive.html

*North American Deserts* http://www.desertusa.com/glossary.html

American Desert Animals: *The Living Desert* http://www.livingdesert.org/

**Chapter 5: The Desert in History**

Blainey, Geoffrey. *The Triumph of the Nomads: A History of Ancient Australia*. London, UK: Macmillan (1982).

Hawkes, Jacquetta. *The First Great Civilizations: Life in Mesopotamia, the Indus Valley and Egypt*. New York, NY: Knopf (1973).

Isaacson, Rupert. *The Healing Land: The Bushmen and the Kalahari Desert*. New York, NY: Grove Press (2003).

*Peoples of the Desert (Native Americans)*. Alexandria, VA: Time-Life Books (1993).

Phillips, E.D. *The Royal Hordes: Nomad Peoples of the Steppes*. London, UK: Thames & Hudson (1965).

Slavin, Kenneth. *The Tuareg*. London, UK: Gentry Books (1973).

Weir, Shelagh. *The Bedouin*. London, UK: British Museum Publications (1990).

White, John Manchip. *The Great American Desert: The Life, History and Landscape of the American Southwest*. London, UK: Allen & Unwin (1977).

**Chapter 6: The Modern Desert**
**Atlas: Australia and the Poles**

Haynes, Roslynn D. *Seeking the Centre: The Australian Desert in Literature, Art and Film*. Cambridge, UK: Cambridge University Press (1999).

Lukas, Mike. *Antarctica the Beautiful*. London, UK: New Holland Publishers (1996).

Moorhead, Alan. *Cooper's Creek*. Harmondsworth, UK: Penguin (1988).

Yelverton, David E. *Antarctica Unveiled*. Boulder, CO: University Press of Colorado (2000).

**Chapter 7: Wealth from the Desert**

Vicker, Ray. *The Kingdom of Oil: The Middle East, its People and its Power*. London, UK: Robert Hale (1975).

Walker, A.S. *Deserts: Geology and Resources* http://pubs.usgs.gov/gip/deserts/

**Chapter 8: Spreading Deserts**

Agadzhan, G. Babaev (ed.). *Desert Problems and Desertification in Central Asia: The Researches of the Desert Institute*. Hamburg and Berlin, Germany: Springer-Verlag (1999).

Balling, Robert C., and others. *Interactions of Desertification and Climate*. Oxford, UK: Oxford University Press (1995).

Bigelow, Paul. *Deserts on the March*. Reprint edn. Washington, DC: Island Press (1988).

Breckle, Siegmar W., and others (eds). *Sustainable Land Use in Deserts*. Hamburg and Berlin, Germany: Springer-Verlag (2001).

Hoekstra, T.W., and Moshe Shachak. *Arid Lands Management*. Champaign, IL: University of Illinois Press (1999).

Middleton, Nick, and David Thomas (eds). *World Atlas of Desertification*. 2nd edn. New York, NY: Edward Arnold (1997).

Reith, Charles C. *Deserts as Dumps: The Disposal of Hazardous Materials in Arid Ecosytems*. Albuquerque, NM: University of New Mexico Press (1992).

*Man-Induced Desertification* http://www.unu.edu/unupress/lecture12.html

United Nations Convention to Combat Desertification http://www.fao.org/desertification/default.asp?lang=en

*Wind Erosion and Dune Stabilization in Ningxia, China* http://www.weru.ksu.edu/symposium/proceedings/mitchell.pdf

# Index

Page numbers in *italics* refer to picture captions.


A
Aborigines 135, 140–1, *149*, 151, *157*, 159, 161, 163
  land rights 155, 161
acacias 77, 78, 81, 83–4
  raddiana 77
  roots 77
  sand 85–6
  symbiosis with ants *80*
adaptations
  animal 89, 90–5, 111
  plant 75–80
adders
  northern death *101*
  saw-scaled 101–2
aeolian landscapes 52
African deserts
  animals 92, 103, 105, *106*, 107–9, 110, 111
  minerals 176–7
  people 132–6, *137*, 138–9
  plants 78, 83–5, 86
  *see also* Arabian Desert; Kalahari Desert; Namib Desert; Sahara Desert
Afro-Arabian Shield 39
agaves 78
"Age of Deserts" *15*, 16
Ahaggar Mountains 27
Air Massif 26
Algeria, resources 176
Alice Springs 160, *163*
alluvial fans and aprons 47
American deserts *see* North American deserts; South American deserts
amphibians 93, 102–3
Amu Dar'ya River 46, 60, 62, 63, 176
Anasazi culture (Pueblos) 142, 149, 151
  Western Anasazi (Sinagua) 119, 121
Andes Mountains, rain shadow 20
Andrews, Roy C. 148, 153–4
androctonus 98
animals 89–111, 183
  polar 23, 165, 167
Antarctica 22–3, *166–7*, 169
  resources 176, 177
Antarctic Circle 166
Antarctic Treaty 167
antelope 35
ant lions 96
Antofagasta 124, 127
ants 96
  symbiosis with acacias *80*
Anzo Borrego Desert *113*
Apaches 143
aquifers *51*, 172, 184
'Arabah, Wadi 46
Arabian Desert 18, 38–40, 44
  animals 40, *92*, *106*
  area 39
  Empty Quarter (Rub' al-Khali) 39, 40–1, 148
  explorations 39, 147–8
  a high-pressure climate desert 39
  plants 40
  resources 176
  Syrian Desert 39, *131*
  *see also* Riyadh
Arab Revolt 136
Aral Sea 60, 183
Aravalli Range 67
Arctic 22, *164–5*
Arctic Circle 164
Arctic Ocean 164
Argentina
  fossils *15*
  *see also* Patagonia
Argentino, Lake *129*
*Argentinosaurus* 15
Arizona
  Anasazi people 142
  Montezuma Castle National Monument 119, *121*
  Petrified Forest National Park *46*
armadillos 106
arroyos 45, *46*
arthropods 90
Ashkhabad 61
Asian deserts
  animals 92, 103, 107, 109–10, 111
  continental deserts 20–1
  minerals 175–6
  people 139–40
  plants 85–6
  Silk Road 61, 69, *138*, 139–40
  *see also* Gobi Desert; Iranian Desert; Kara-Kum Desert; Taklimakan Desert; Thar Desert
asses, wild 72–3, 107
Atacama Desert 18, 79, *124–7*
  animals 102, 106, 107, *125*, 126
  climate 125
  mining/minerals 124, 125, *169*, 172, 176
  people 127, 143
  plants 79, 83
Atacama Salt Flat 124
Atacameño 127
Atlas-Erg *29*
Atlas Mountains 28, 29
atmospheric pressure, sub-tropical high-pressure deserts 16–18
Australian deserts 18, 21, 162–3
  animals 92, 93, 95, 99–106, 109–10, 111, 159, *162*, 163, 183
  desertification *179*
  dunes 56
  explorations 148, *149*, 150
  gibbers 51
  modern crossings 154, 161
  national parks 158, 159, 161
  people *see* Aborigines
  plants 78, 80, 86–7, 159, 163
  resources 159, 161, 176, 177
  *see also* Central Australian Desert; Western Australian Desert
Ayers Rock *see* Uluru
azorellas, three-lobed 83

B
Badain Jaran Desert 55–6, 70
Badwater 114
bahadas 48
Baja California 119
bajadas 47
Bam 65
bandicoots 111
bandy-bandy 102
baobabs 77–8
barchan dunes 37, 54–5, 123
Barsa Kel'mes ("Place of No Return") 45, 62
Barth, Heinrich 147
basalts 14
bats 93
Bayóvar Depression 123
Béchar (Colomb) 29
bedouins 40, 135, 136–7, *150*, 151
beetles 33, 89, 97
Benguela Current 17, *18*
Berbers 133–4
Bering Strait 165
Betty Ford Center 121
bhakars 66
bindweeds 85
biomass, standing 77
biomes, major 77, *81*
birds 103–6
  polar 22, 23, 167
Biskra 29
blackboys (grass trees) 86
"black racers" 102
Black Rock Desert 115, 116
Blunt, W.S. 148
bobcats 109
Bonaparte, Napoléon 146
Bonneville, Lake 115
borax 175
Botswana 34, 35
browningias 78–9
budgerigars 104
Bukhara 131, 140
Buraydah 39
Burckhardt, J.L. 147
Burke and Wills expedition 148, 150
burrows 91, 92
Burton, Richard 147–8
buthus, Saharan 98
buttes 52

C
cacti 81–2, 83, 119, *121*
  elephant 81
  opuntia 87
  Queen of the Night 77
  roots 77
  saguaro 81, 119, *121*
  spines 79
  water storage 78
Caillé, René 135, 146–7
California *see* Death Valley; Great Basin; Mojave Desert; Sonoran Desert
camels *63*, 134, *135*, 137
  racing 151
Canning stock route 159
caracals 109
caravans 134, 139
carnivores 89–90, 107–11
cats, sand 92, 109
cedar trees, pygmy 77
centipedes *95*
Central Australian Desert *160–1*, *162*
  Coober Pedy 161, 177
  Uluru 141, *149*, 161
Central Kalahari Game Reserve 35
Chang Ch'ien 139
cheetahs 107–9
Chile, deserts *see* Atacama Desert
Chimú Empire 173
China, deserts *see* Gobi Desert; Taklimakan Desert
Chinchorro people 127
chotts 29
Chubut River 128
Chuquicamata 124, *169*
civilizations, ancient 131, 132–3
Clapperton, Hugh 146
cliff dwellings 119, *121*
  Cliff Palace *142*
climate change 185
coachwhips 102
coastal deserts 17, 33, 123, 125, 129
Coastal Ranges 20
*Coelurosaur 15*
Colorado, Mesa Verde 142
Colorado Desert 118, 120
Colorado River 45, 119
Comodoro Rivadavia 129
Comstock Lode 115, 116
Conchas, Las 127
conifers 75
continental deserts 20–1
Convention to Combat Desertification 179
Coober Pedy 161, 177
Coolgardie 159
Copiapó 124
Coriolis effect 16
Costa, Cordillera de la 124
coyotes 109, 110
cram-cram 85
creosote bushes 77, 80, 81, 82
crops, genetically modfied 173–4
crustaceans 94
currents, ocean 17
  *see also* Humboldt Current

D
Dalandzadgad 71
Death Valley *13*, 20, 114, 116, *117*
  alluvial fan *47*
  minerals 175
  name 116
degradation and erosion 182–3
  *see also* desertification; weathering and erosion
delta, dry *13*
desalination 153
Deseado River 129
Desert Fathers 136
desertification 179, 180–185
  in the Taklimakan Desert 69
desertization 180
desert peas, Sturt's 80
desert varnish 26, 47
detrivores 90
Devil's Marbles 161
dew 33, 97, 125
dhands 66
diamonds 32, 35, 37, 159, 165
dingoes 109–10
dinosaurs 15, 128, 167
Djerid, Chott 29
dogs 109–10
Domeyko, Cordillera 124
Doughty, C.M. 148
dragons, bearded 100–1
dress 135
dromedaries 137
drought 80
Dumont d'Urville Sea 167
dunes 28, 29, 33, 52–6, 81
Dust Bowl *184*, 185
"dust devils" 56
dust storms 56–7
Dzungarian Desert 72

E
eagles, wedge-tailed 105–6
Egypt 146, 174, 176
  resources 176–7
  *see also* Western Desert
Egyptians, ancient 132–3, 169
El Alamein 30
El Niño 17, 122, 123
El Tatio Geysers 125
Empty Quarter (Rub' al-Khali) 39, 40–1, 148
emus 104, *162*
energy 174
  solar 169, 174, *175*
ephemeral plants 80, 83, 120
erg, Great Erg of Bilma 54
ergs 28, 29, 44
erosion *see* weathering and erosion
estivation 93
Etosha Pan 32
Etosha Park 32
euphorbias 77, 78, 79–80, 83
Euphrates River 50, *133*
evaporation, water 36
evening primroses, white dune 75
explorations, European 146–50
Eyre, Lake 161, *162*, 163

F
falcons 105
farming 153, 173–4, 182
fennecs *110*
fennels, giant 86
fertilizers, from deserts 172
Finke Gorge National Park 161
Fire, Valley of *77*
fires 86
fish 94
Fitzroy River 158
flamingoes *125*
flannel flowers 86–7
floods 26, 48–50
flowers 75, 85, 86–7
fogs, coastal 33, 97, 123
food chains 89–90
formation, desert 13–21
Forrest, John and Alexander 150
fossils 15, 128, 167
foxes 107–8, 110–11
Franz Josef Land 165
Fremont, John 149
frogs 93, 103, 159
fronts, weather 18

G
Gaborone 35
Gansu Desert 72
*garúa* 123
gazelles 93, 107
gemsbok 35
Gemsbok National Park 35
Genghis Khan *139*, 140
gerbils, Mongolian 106
Ghadamis 29
Ghan Railroad 161
Ghansiland 35
Ghardaïa 29
gibbers 51
Gibson Desert 21, 56, 158–9, 177
gila monsters 101
Giles, Ernest 150
glaciation 164–5
glaciers
  Arctic 165
  melting 23
glass, desert 30
global warming 23, 185
goannas 99
goats 182
Gobi Desert 15, *19*, 21, 70–3
  animals 72–3, 93, 102, 106
  explorations 148, 154

fossils 15
modern crossings 71, 154
name 72
plants 72, 80, 86
population 72, 73
rainfall 72
resources 176
temperatures 44
Trans-Mongolian Railroad 71, 73
weathering and erosion in 45
gold 115, 116, 134–5, 159, 161, 169
Gold Rush 116
golf courses *153*, *182*
goosefoots 85
Gosse, William C. *149*
Grand Canyon 115, 116
grasses 81, 84–5, 87
pampas 83
grass trees 86
grazing, by livestock 181–182
Great Basin *18*, *114*–17, 175
animals 115
area 115
climate 115, 117
formation 117
people 116
plants 115, 116
resources 176
*see also* Death Valley; Mojave Desert
Great Basin National Park 115
Great Dividing Range 21
Great Eastern Erg 29, 55
Great Erg of Bilma 54
Great Indian Desert *see* Thar Desert
Great Man-made River 30
Great Salt Desert (Dasht-e-Kavir; Kavir Desert) 64, 65
Great Salt Lake 115, 116
Great Salt Lake Desert 115, 116
Great Sandy Desert 44, 158, 159, 177
Great Victoria Desert 159, 160, 177
Great Western Erg 29
greening the desert 172–4
*see also* irrigation
Greenland Glacier 165
groundwater 45, 173, 184–185
guanacos 107
Gulf War (1991) 169, 171
gum trees 86, *159*
Gurvan Sayhan 71
gypsum 49

H
Haber–Bosch process 172
haboob 56
Hadley Cells 18
Ha'il 39
Hajr, Wadi 39
Halls Creek 158
hamadas (hammadas) 28, 44
hedgehogs, desert *106*
Hedin, Sven 148
hematite 47
herbivores 89
mammalian 106–7
repelled by plants 79–80
hermits 136
*Herrerasaurus* 15
Himalaya mountains, rain-shadow *19*, 21
history 130–43
Hohokam Indians 121
Hornemann, F.K. 146
horses, Przewalski's 73
Hotan 69
Humboldt Current 122, 124, 125, 126
Humboldt River 115
humus 44
Husayn ibn 'Ali 136, 137
Hussein, Saddam 152–3, 171
hyacinths 86
Hyderabad 67
hyenas 107–8, 111

I
ice, causing weathering 45
Idaho, desert *see* Great Basin
Imperial Valley 118
India, desert *see* Thar Desert
Indira Ghandi Canal 67
Indus River 67
insects 33, 89, 92–7
inselbergs 33
Iran
minerals 176
Plateau of 64
Iranian Desert *64*–5
Dasht-e-Kavir 64, 65
Dasht-e-Lūt 53, 64, 65
rainfall 65
salt marshes 64, 65
Iraq 171, 175, 176
ironwoods 86
irrigation 30, *153*, 169, 172–3, 174, 183
Isa, Mount 161
Israel
irrigation *173*
*see also* Negev Desert

J
jackals 111
Jāfūrah, Al- 40
Jaisalmer 67
Jaladah, Al- 40
Jericho 131
Jerome, St. 136
Jodhpur 67
Jordan 175
Joshua Tree National Monument 118
Joshua trees 82, 118, 119, *121*
Jurassic period 15
Jwaneng 35

K
Kachichh, Rann of 67
Kalahari Desert 15, 18, 34–7, *56*
animals 35, 37, 101, 111
area 36
explorations 147
national parks and game reserves 35
Okovango Delta 34, 37, 50
people 35, 37, *137*, 138–9
plants 35, 37, 79, 83–4
rainfall 36
resources 35, 37, 176, 177
rock painting from *137*
salt pans 35, 48, 94
temperatures 36
Kalgoorlie 159
kangaroos 111, 161
Karakorum 71
Kara-Kum Canal 61, 63
Kara-Kum Desert 20–1, 44–5, 46, *60*–3
animals 63
area 63
Barsa Kel'mes ("Place of No Return") 45, 62
climate 63
formation 63
minerals 63
name 63
plants 85–6
population 63
rainfall 63
sands 62
takyrs 52
Kavir, Dasht-e- (Kavir Desert; Great Salt Desert) 64, 65
Kavir Buzūrg 64
Kavir National Park 64
kavirs 64, 65
Kazakhstan 176
Kenya, desert management 181
Kermān 65
Kerulen River 45
Khauz-Khan reservoir 61
Khoi people 33
Kimberley Plateau 158
Kitt Peak Observatory 119
Kopet-Dag Oasis 61
kopjes 36
Koussi, Emi (Mount Koussi) 26, 28
Kubango River 50
Kufrah, Al- 30
Kunlun Shan 68
Kutse 35
Kuwait *176*
Kyzyl-Kum Desert 60, 62

L
Laghouat 29
Laing, Alexander 146
lakes 45, 48
Lambayeque 123
lanceheads, snouted 102
languages, click 138
"La Niña, Lake" 122
larkspur 86
"Last Frontier, the" 162
Las Vegas 115
Lehman Caves 115
Leichardt, Ludwig 149–50
Lena River 165
Letihakane 35
liaretas 83
Libya, resources 176
Libyan Desert 30–1, *132*
limestones 14
*Lithops* spp. *84*, 85
Livingstone, David 34, 147
"living stones" *84*, 85
lizards 90–1, 99–101
monitor 99
Llullaillaco 124
Loa River 125
locusts, desert 94–5
*lomas* 123
Lop Nur 68
Los Angeles 119
Luna, Valle de la 124, *127*
Luni River 67
Lūt, Dasht-e- 53, 64, 65

M
Macdonnell Ranges 160
Magellan, Ferdinand 129
Magellan, Strait of 128
Makgadikgadi Pans 35, 48
*makhteshim* 46
Mali, resources 176
mammals 90, 91–2, 106–11
Mandalgovi 71
manushis 102
Marble Bar 158
margays 109
Ma'rib 39
marsupials 95, 111
Marusthali Desert 67
Mary 61, 63
Mauritania, resources 176
Mecca 39, 147
Medina 39
mesas 52
Mesa Verde 142
Mesopotamia 131, 132
mesquites 77
metals and minerals 171–2
*see also* minerals/mining
meteorites 167
Middle Eastern deserts
access to water 174
animals 91, 100–2, 107, 109
oil 152–3, *170*, 171
people 136–7, 151
*see also* Arabian Desert
migration 93
minerals/mining 35, 37, 124, 125, 161, *169*, 171–2, 175–7
mirages 57
mites 97, 98
Mogollons 142
Mojave Desert *18*, 19, 94, 118–19, *121*, 175
area 119
birds 104
climate 119
minerals 119, 176
plants 119, *121*
moles, marsupial 111
molochs 100
Mongolians (Mongols) 59, 73, *139*, 140, *154*
monks 136
Montezuma Castle National Monument 119, *121*
morama bulbs 84
Mormons 116
Morocco, resources 176
moths, yucca 83
"mud volcanoes" 45
mulgas 78
mummies and mummification 127, 132
Murgab Oasis 61
Murgab River 61
Musa, Mansa 135
Mu Us Desert 70

N
Nabi Shu'ayb, Mount An- 39
Nafūd, An (Great Nafud) 39
Namib Desert 17, 18, 32–3, *43*, 44
animals 33, 97, 101, 111
a cold-water coastal desert 17, *18*, 33
dunes 33, *54*
floods *48*
fogs and dew 33, 97
name 33
national parks 32
people 33
plants 33, *79*, 81, 84
rainfall 33
resources 176, 177
sands 33
temperatures 33
Namib-Naukluft Park 32
Napoléon Bonaparte 146
Native Americans *81*, 142–3, 151–2, 154–5
natural gas 63
Navajo 143, 151, 152, *155*
Nefta 51
Negev Desert *150*, 153
irrigation 173, 174
*makhteshim* 46
Negro River 128
Neuquén 128
Nevada
mining 116
Valley of Fire *77*
*see also* Great Basin
New Mexico
fossils *15*
pueblos 142, *143*
White Sands National Monument 49
New Valley Project 30, 174
Ngami, Lake 34
nickel 159, 161
Niger, resources 176
Niger River 45
Nile River 27, 45, 143
nitrates 172
Nitria 136
nomadic people 131–2, 151, 152
*Nomad* (robot) 125
North American deserts 19–20, 52
animals 91, 95, 101, 102, 104, 107, 109, 110–11
explored 149
irrigation *153*
minerals 175
people 142–3, 151–2, 154–5
plants 77, 80, *81*–3
rain-shadow deserts 16, 19–20
wind-formed rock formations 52
*see also* Dust Bowl; Great Basin; Mojave Desert; Sonoran Desert
Nubian Desert, resources 176–7
Nullabor Plain 154
Núñez Cabeza de Vaca, Alvar 149

O
oases 50–1, 52
Ob' River 165
oil, crude 152–3, 169, 170–1
from the Arabian Desert 39, 40
from the Arctic 165
from the Iranian Desert 65
from the Kara-Kum Desert 61
from the Libyan Desert 30
from the Taklimakan Desert 69
wars over 171
Okovango Delta 34, 37, 50
Olga, Mount 161
opals 177
Oranjemund 32
Organ-pipe Cactus National Monument 118
orographic lifting 19
oryxes 107
ostriches 103–4
Oswell, W.C. 147
owls, elf 104

P
Pacific, War of the 125, 172
Pahlavi, Mohammed Reza 152
paintings, rock *137*, *141*
Pakistan, desert *see* Thar Desert
Palgrave, William 148
Palm Springs 118, 120, 121
palm trees, Livingstone 86
Palmyra 39, 131
Pamirs 68
*pamperos* 20
Pan-American Highway *153*
Panamint Mountains 20
Pangaea 13–14, 15
pans, salt 34, 35, 39, 48
Parque Nacional Los Glaciares 128

Parque Nacional Perito Morena 128
Patagonia 15, 19, 20, 46–7, 106, *128–9*
  animals 107, 129
  climate 129
  name 129
  national parks 128
  people 129, 143
  plants 83, 129
  rivers 46–7
Patagonian ice sheet 128
pediments 47
peludos 106
penguins, Adélie 22
people 131–43, 151–2, 154–5
perenties 99
Permian period 13–14
Peru, deserts *see* Sechura Desert
Petra 131
petrels, snow 22
Petrified Forest National Park *46*
petroleum *see* oil, crude
Philby, Harry St. John 148
phosphates 123
photosynthesis, CAM 78
phyllodes 78
pigs 182
pines, bristlecone 82–3, *117*
pipelines
  oil 61
  water 30, 31, 32
pistachio trees 85, 86
Piura 122
Piura River 123
"Place of No Return" 45, 62
plants 22, 74–87
  preventing desertification 181–2
  *see also* crops
playas 48
  *see also* salt flats/pans/lakes
poikilotherms 90
polar deserts 22–3, 165
  mirages in 57
Polo, Marco 72, 140
poppies 85
"Porcelain Desert" (White Sands National
Monument) 49
predators 107–9
prehistoric deserts 13–15
pressure, atmospheric *see* atmospheric pressure
prickly pears 87
productivity, primary 77
pronghorns 107
Pueblo Indians 154
pueblos 142, *143*
Purnululu National Park 158
pyramids 30, 132

**Q**
qanats 173
Qattara Depression 27, 28
Qiemo 68
Queen of the Night cactus 77
quiver trees 78, 79

**R**
rabbits 183
railroads 71, 73, 154, 161
Rainbow Bridge 47
rain forests *17*
rain-shadow deserts 18–20, 117
rats, kangaroo *111*
rattlesnakes, sidewinding 91
Rawson 129
"red racers" 102
regs (desert pavements) 28, 44
reptiles 90, 92, 99–102
resources 175–7
  *see also* minerals/mining; oil, crude
rheas 126
ringtails 111
Rio Negro (province) 128
rivers
  dry channels 45, 46
  ephemeral *43*, 45–6
Riyadh 39, *145*, 153
roadrunners 104
roads, Pan-American Highway *153*
rocks 14–15, 44
  formations 47, 52, 53
  *see also* weathering and erosion
rodents 106
roots, plant 76, 77
Ross Sea and Ice Shelf 167
Route 40 128
Rub' al-Khali (Empty Quarter) 39, 40–1, 148
Rum, Wadi *14*

**S**
Sab'atayn, As- 39
safety, when traveling 152
Safsaf Oasis 52
sagebrush 81, 82, 115
Saguaro National Park 119
Sahara Desert 18, 26–8, 44, 53
  animals 92, 95, 101–2, 103, 107, 109
  Atlas-Erg 29
  climate 28
  dunes 28, 29, 54
  during the last ice age 180
  dust storms 56–7
  explorations 146–7, *148*
  Great Erg of Bilma 54
  a high-pressure desert 28
  Libyan Desert 30–1, *132*
  modern crossings 153
  name 28
  oases 51
  people 28, 133–6, 151
  plants 27, 84–5
  resources 176
  rivers 27, 29, 30
  Sahel 184
  shifting 180, *181*
  size 28
  temperatures 27
  Ténéré Desert 25, 27, 55
  terrain 28
  trade routes 28, 134–5
Sahel 184
salars 48
  *see also* salt flats/pans/lakes
salt 134, *172*
salt basins, largest 35, *48*
salt flats/pans/lakes 39, 40, 45, 48, 51, 94, 124, *127*
Salt Lake City 116
salt marshes (kavirs) 64, 65
Salton Sea 118
Salt River 119
Samarqand 131, 140
San 35, *37*, 135, *137*, 151
  languages 138
sand 44, 52–7
  gypsum 49
sandfish 100–1
sandgrouse 105
sandstone 14, 15, 52
sandstorms 56–7
sand verbena, pink 75
San Pedro de Atacama 125
Santa Cruz 128
satellite photography 52
Saudi Arabia 39, *145*, 151, 152, 153, 175
saxual trees 85
scarabs 97
scorpions 97–8
seafog *see* fogs, coastal
sebkhas 39, 40
Sechura Desert 18, 78–9, *122–3*, 172
  ancient people 123
  animals 107, 123
  climate 123
  a cold-water coastal desert 123
  dunes 123
  minerals 123, 175, 176
  Pan-American Highway *153*
  plants 78–9, 83, 123
secretary birds 105
seif dunes 54
selenite 177
Sequoia National Park 114
Shawia people 29
sheet wash 47
shepherd's trees 84
shrimps 94
Sierra Nevada 20, 114, 115, 116
  rain shadow *13*, 114
Silk Road 61, 69, 131, *138*, 139–40
Simpson Desert 21, 161
Sinagua Indians 119, 121
Siwa 30
Skeleton Coast 32, 33
skinks 99–100
slaves 135–6
Smith, Jebediah 149
snakes 33, 91, 101–2
soils 44, 182–183
solifuges (sun spiders) 91, 98
Sonoran Desert 118–21
  animals 104, 119
  climate 120–1
  a green desert 120
  name 119
  people 121
  plants *113*, 119, 120
Sossusvlei 32, 33
South American deserts 20
  animals 104, 107, 109, 126
  explorations 149
  minerals 175
  people 143
  plants 81, 83
  rain-shadow deserts 16, 20, 126
  *see also* Atacama Desert; Patagonia; Sechura Desert
Southern Ocean 167
south pole 167
space, deserts seen from 52
spinifex 87
spreading deserts 179–85
spurges 83
squirrels, ground 92
streams, exotic 45
stringers 56
Stuart, John McDouall 150
Stuart Highway 161
Sturt's Stony Desert 51
Subtropical Convergence 167
subtropics, high-pressure deserts 16–18
succulent plants *see* agaves; cacti; euphorbias
Sudan, resources 176–7
sun spiders (solifuges) 91, 98
supercontinent, Pangaea 13–14, 15
symbiosis 80, 83
Syrian Desert 39, *131*

**T**
Tademait Plateau 26
Taklimakan Desert *19*, 21, 68–9, 95, *138*
  resources 176
takyrs 52, 62
Tamanrasset 26
tamarisks 85
Tamerza *51*
Tanami Desert 161
Tarim River 45, 68
Tashk, Lake 65
Tehran 64
Tehuelches 143
temperatures
  affecting weathering 44–5
  animals and 90, 91, 92–3
Ténéré Desert 25, 27, 55
terminalia trees 86
termites 92–3
Thar Desert (Great Indian Desert) 16, 66–7, 176
  resources 176
Thesiger, Wilfred 148
Thomas, Bertram 39, 148
thorny devils 100
Tian Shan 68
Tierra del Fuego 129
Tigris–Euphrates River 45
toads 102–3
Tocopilla 125
Tombouctou (Timbuktu) 26, 134, 135, 146
Tom Price, Mount 177
torpor 93
tortoises 99
Toubkal, Mount 29
tourism 153
Trans-Altay Gobi 70
Trans-Mongolian Railroad 71, 73
transportation 152, 153–4
trees 181, 182
Triassic period 14–15, 15
tropics 17–18
Tuareg 26, 133–4, 146, 151
  dress 135
Tularosa Basin 49
tulips 86
Turkmen 63
Turkmenistan
  minerals 176
  natural gas 63
  *see also* Kara-Kum Desert
Turpan Depression 68

**U**
Ulaanbaatar 71
Uluru 141, *149*, 161
United Arab Emirates *182*
Ur *133*
Utah
  Anasazi people 142
  Rainbow Bridge 47
  Salt Lake City 116
  *see also* Great Basin
Uyuni Salt Flat 125

**V**
Valdés Peninsula 129
Valley of the Dead 20
Valley of Fire 77
Varthema, Ludovico di 147
Vatna Glacier 165
*Velociraptor* 15
vicuñas 107, 126
Viedma, Lake 128
Vinson Massif 167
vipers
  Asiatic pit 102
  saw-scaled 101–2
  sidewinding *91*, 101
Vizcaino Desert 119
volcanoes, in Antarctica 167
vultures, turkey 106

**W**
wadis *14*, 27, 39, 45, 46
Walker, Joseph 149
Warburton 158
Warburton, Peter 150
water
  animals and 90, 91, 96
  aquifers *51*, 172, 184
  desalination 153
  effects 45–7
  groundwater 45, 173, 184–185
  metabolic 91
  mismanagement 183–184
  pipelines 30, 31, 32
  plants and 76, 77–8
  surface water 45
  wars over 174
  weathering and erosion by 46–7, 48–51
  *see also* irrigation
wealth from deserts 169–77
  *see also* gold; minerals/mining; oil
weathering and erosion 43–4, 182–183
  by water 48–51, 445–7
  by wind 28, 51–2, 53, *127*
  effects of heat and cold 44–5
  onion-skin weathering 45
  plants prevent 182
wells, Great Man-made River and 30
welwitschia 33, 77, 84
Western Australian Desert 158–9
  Gibson Desert 21, 56, 158–9, 177
  Great Sandy Desert 44, 158, 159, 177
  Great Victoria Desert 159, 160, 177
Western Desert, irrigation 30
whirlwinds 56
White Desert 23
White Sands National Monument 49
Whitney, Mount 114
"willy willies" 56, 159
winds 16, 17–18
  erosion by 28, 51–2, 53, *127*
  monsoon 66
  moving sand 52–7
  and rain-shadows 18–20
  and sandstorms 56–7
  trade 16
wind spiders *see* sun spiders
wood 181, 182
Wyoming, desert *see* Great Basin

**Y**
yams 83, 84
yardangs 52, 64–5
Yelyn Valley National Park 71
yuccas 83
Yuma Desert 119, 120

**Z**
Zaunguzk Plateau 46
Zeil, Mount 160
zeugen (pedestals) 52